Climate Futures

Climate Futures

Time Travels to Alternative Outcomes

Jack Kozuchowski with Brian Epp

Copyright 2016 Jack Kozuchowski with Brian Epp
All rights reserved.

ISBN: 1536871044
ISBN 13: 9781536871043
Library of Congress Control Number: 2016912580
CreateSpace Independent Publishing Platform
North Charleston, South Carolina

DEDICATION

This book is dedicated to Gaia.

Who is Gaia? In Greek mythology, Gaia was the great mother of all. She was the primal Greek mother goddess who created and gave birth to the earth and the universe.

In modern times (the 1970s), this concept of "Mother Earth" was used by James Lovelock, a chemist, and Lynn Margulis, a microbiologist, to create a theory of all life and matter integrated as one entity. The Gaia Theory holds that living organisms interact with the non-living elements of the earth to establish a synergistic, self-regulating "super-organism." This involves all matter of our planet, acting as one to maintain the conditions necessary for life.

It is birth. It is death. It is the recycling of our bodies back into the earth. It is the stream of consciousness that connects all living beings on the planet. It is part of us and we are part of it. The earth is our foundation and we are its stewards.

Let us treat the earth well.

DISCLAIMERS

Some of the characters in this book are named after real people – with a slight modification of their names. This device was used by the principal author to make these characters more real to him. Readers of this book who happen to know people with real names that are strikingly close to the names of the characters in the book should recognize that the personalities and perspectives of these characters are fictional and do not necessarily reflect the personalities and/or views of the real people.

All images used in this document are public domain or have received permission from its owner. Note that the image of Gaia on the previous page was downloaded from the Internet on August 18, 2016 with implicit permission from its owner, which states, on the website that backs the image: *"This file has been released to the public domain by the copyright holder, its' copyright expired or it is ineligible for copyright. This applies worldwide."*

See: http://riordan.wikia.com/wiki/File:Presence_of_gaia.jpg

This project has been funded by and falls under the aegis of Kozuchowski Environmental Consulting, LLC.

PRINCIPAL AUTHOR'S FOREWORD

To my readers,

Welcome to the future! That is a phrase that you can use at the end of each day, each hour, each second. Indeed, the future is always one blink away as the present moment fades. Yet obviously, we do not sense entering into the future at this micro scale. To really see and feel the future, we need our imagination and the convention of time travel. So, as you read this book, release your imagination and fly forward into the different scenarios to see how we, as a society, are doing.

First, I offer a disclosure of transparency. My co-author and I are firmly entrenched in the position that climate change is real, caused by humans, and will result in a future world that will be unpleasant to say the least . . . unless we mitigate the problem. However, when writing this book, we set aside our passion toward this conviction to envision many possible outcomes that could occur.

This is a book of fiction—a story. It is also fiction that is based on plausible science. As such, it will allow you to learn something. You will learn about climate change, whether it be the evidence of warming, the level of certainty required to say that climate change is occurring, exploring the various ways to fix the problem, and the overall picture of what different futures may be, depending on what we do or do not do to fix the problem today. The science will come through to you, even if you skim over the technical discussions of panels that occur periodically throughout the book. If you want to dig deeper, all technical points that are based on today's plausible science are footnoted, so you can read more about the details.

But there are other lessons in this story as well, such as how a genuinely open mind can enrich our perspectives, how we can negotiate through conflicts to create win-win outcomes, and when to "Let it Be." (Note that the phrase throughout the book is capitalized, as it was inspired by the Beatles song by the same name.)

The theme of this book is the issue of climate change and what we may do (or not do) to reduce the levels of heat-trapping chemicals, such as carbon dioxide and other greenhouse gases. To keep it simple, we will refer to all heat-trapping gases as carbon dioxide—CO_2. Just know that there are many other gases that trap heat (e.g., methane, nitrous oxide, ozone, chlorofluorocarbons, and sulphur hexafluoride).

Our central character, Dr. Mitch Warner, is a climate ecologist. In Chapter 1, we are in his present-day world of 2020 (I could have just as easily called the present 2016, but in a few years, that would have been dated.) So just assume that 2020 is the present moment. See, you have already become a futurist!

Chapters 2–4 begin the time travel to Dr. Mitch's future. Whenever he enters a door, he is entering a different future scenario, stemming from what we do about carbon dioxide (CO_2) emissions. Beyond each door, he jumps to different time periods: same channel, different time. So the story evolves on the premise that society is at a crossroads as to whether to take action at the present moment. The future jumps to time periods that evolve from the consequences of that action (or in the case of Door #1, no action). As the conflicts and consequences are resolved or not resolved, we enter the next door.

At the end of the journey through each door, Mitch absorbs what he has learned, then returns to the vestibule and enters the next door. He is then carried forward to different time periods within the door he is in, with different assumptions about what we do (or don't do) to deal with climate change: same channel, different times. In an attempt to make these time jumps visually comprehensible, the Table of Contents below illustrates where Mitch will be traveling.

Chapters 2–4 also describe scenarios that are based on what the United States (and other world governments) do, which will dictate the consequences of that time. The three doors present scenarios that are generally used by today's climate modelers: we do nothing (Business as Usual—Door #1), we do a modest amount to address climate change (Door #2), or we give our full effort to mitigate emissions (Door #3). These chapters are relatively short. You will quickly recognize the consequences of each of these scenarios, because this is what scientists have been currently telling us. Yet, it is important to start with this foundation of understanding. For those of you who are sick of the gloom and doom stories such as Door #1, bear with me, as a more optimistic outcome is presented behind Door #3 and the "thinking outside of the box" outcomes are presented in the remainder of the book.

Doors #4–7 provide different perspectives. Through door #4, the skeptics approach is adopted, with the attitude: "Before we act, prove to me that climate change is happening beyond any doubt." Chapter 5 proposes a world where geo-engineering is evaluated and deployed. Can we tinker with the sun's radiation/energy/heat to counter the warming caused by our CO_2 emissions? Stay tuned. Door #6 takes us to the world after the eruption of the Yellowstone supervolcano. What does a world of a continuous winter have to do with global warming? Hint: if society recovers from the cooling affects of the blast, will it re-industrialize with lessons learned from the first industrial revolution.

Chapter 7 addresses an unconventional approach to solving climate change from the bottom up without government. This is the "grassroots" scenario.

A convention in the book bears explanation. The "gold standard of mitigation" that we use in this novel is to achieve 80 percent reduction of emissions relative to 1990 levels of CO_2. This standard means "enforceable and/or attainable" goals. A special note on what "enforceable and/or attainable" means in terms of the Paris climate treaty of 2015 is discussed in the note at the end of this novel. Suffice it to say that the Paris Agreement will not, by itself, bring us to 80 percent emission cuts.

Another convention: italicized sentences represent the main character's thoughts. You will note that in these thoughts, the voice shifts to the first person (Also note that when individual words are italicized, it is simply for emphasis).

Finally, I need to note the panels, juries, and conferences. This is the main vehicle that we use to convey scientific information about climate change. It may take some extra concentration to plow through this material, but doing so will enrich your experience in reading the book. When you reach each of the five discussions and debates, put on your "learning hat," slow down a bit, and pay attention. It will be worth your while. After the Table of Contents you will find a list of the major characters and panelists following the timeline.

I truly hope that you enjoy this journey into different futures. I hope that, beyond being entertained, you learn something from these journeys and realize that the fate of the future of the world is in our hands. Which door should we enter? I leave it to you to decide.

Jack Kozuchowski

CO-AUTHOR'S FOREWORD

The collaborative writing between us has truly been a gift. It seems like only yesterday that Jack gave me his first short story draft for review. From there, we embellished it to a point where we feel satisfied with its content and verve. Is a book ever finished? There are endless tweaks and stories that could enliven the reader even more! I thoroughly enjoyed getting into each of the characters' personae, becoming their thoughts and feelings, and embellishing their lives. I hope you as the reader do also.

Jack and I believe there is validity to the theory of climate change and hope you as the reader will view this book as a primer to the possibilities that may, or may not, occur. Ultimately, nature will run its course no matter what we do, although to greater or lesser degrees of severity.

As I embellished Jack's story with anecdotes I felt my own humanness in questioning the "what if" scenarios of the future. Needless to say, the predictions from Nostradamus, the Bible, Edgar Cayce, the Mayan calendar, Y2K, countless movies, and hundreds of other commentaries have all preached doom and gloom across the millennia. And thankfully, we seem to have survived so far. However, alternate universes and a world group consciousness, so to speak, may have allowed us, as a species, to avert possible disasters. Purely conjecture on my part, but it does give me peace of mind as to the possibility that yes; we can alter our fate as a fragile human species living here on our only world. And for me, I hope this book helps to consider our thoughts and dreams as a world, a species, and a collective to all the creatures great and small living on this, our Earth.

Brian Epp

TABLE OF CONTENTS

CHAPTER	DOOR #	Time Periods	Page #
Prologue	Prologue The Reality of the Present: 2020	2020	1
1	The Red Door (#1): Business as Usual	2032 2040 2065	7
2	The Yellow Door (#2): The 20 Percent Solution	2034 2040 2044	20
3	The Green Door (#3): The 80 Percent Solution	2044 2033 2152	26
Interlude	Interlude: 2020	2020	34
4	The Gray Door (#4): The Skeptics	2032 2042 2162	45
5	The Silver Door (#5) : Geoengineering	2032 2034 2162	73

Climate Futures

6	The Crimson Door (#6): The Volcano	2024 2044 2162 2224	109
Interlude	Interlude 2	2020	154
7	The Blue Door (# 7): The Grassroots Solution	2034 2152	161
Epilogue	Epilogue	2020	215

MAIN CHARACTERS

Dr. Mitch: Our main character—a climate ecologist who visits all of the dimensions and timelines exhibited in the Table of Contents

Leah: Dr. Mitch's wife

Susannah: Dr. Mitch's daughter

Leah-Ann: Dr. Mitch's great-times-six granddaughter from Doors # 4, 5, 6, and 7

Lonnie: Dr. Mitch's secretary

Casey: Time/dimension travel guide for Doors #1–3

Jake and Janey: Time/dimension travel guides for Doors #4–6

Reverend Tyndall: Facilitator for Environmental/Engineering Jury for Door #5, and time/dimension travel guide for Door #7

Gabriella: Time/dimension travel guide for Door #7

PANEL/CABINET/ JURY MEMBERS

Chapter #	Door # /Year	Conference / Panel	Participants	Role
1	Door # 1: The Red Door 2044	Panel Discussion	Hanaford Curtiss Bertrand Marcus Harley	Moderator Economist Ecological Analyst Geo-engineer Transportation Specialist
2	Door # 2: The Yellow Door 2044	Panel Discussion	Hanaford Curtiss Harley Mitch Bertrand	Moderator Economics Policy Analyst Economist Climate Ecologist Air Pollution Engineer

Climate Futures

Chapter #	Door # /Year	Conference / Panel	Participants	Role
3	Door # 3 The Green Door 2032	President's Cabinet	President Rother Bertrand Dr. Mitch	President Transportation Secretary Science Advisor
4	Door # 4 The Gray Door 2032	Televised Panel Debate	Hanaford Bertrand Shelski Curtiss Marcus Mitch	Moderator Climatologist Environmental Policy Analyst Economist Geo-engineer Climate Ecologist
5	Door # 5 The Silver Door 2036	Science Jury for Geo-engineering	Reverend Tyndall Che-Ling Hanaford Harley Mitch	Moderator Geophysicist Environmental Engineer Hydrologist Climate Ecologist

Panel/Cabinet/Jury Members

Chapter #	Door # /Year	Conference / Panel	Participants	Role
6	Door # 6 The Crimson Door 2152	Environmental and Engineering Panel for the Re-industrialization of Society	Mitch Che-Ling Harley Bertrand Hanaford Curtiss	Moderator / Climate Ecologist Hydrogeologist Chemical Engineer Transportation Specialist Agronomist Civil Engineer
6	Door #6 The Crimson Door The Volcano 2224	Meeting of Emperor's Advisors	Erik II Che-Ling Angelica Drew Tetra Hanaford	Emperor of the North American Kingdom Steward of the Economies Emperor's Science Advisor Assistant to the Science Advisor Minister of Manufacturing Barrister of External Affairs

Climate Futures

Chapter #	Door # /Year	Conference / Panel	Participants	Role
7	Door # 7: The Blue Door 2034	The Grassroots Coordinating Council Science and Engineering Panel	Mitch	Climatologist / Moderator
			Che-Ling	Geophysicist
			Harley	Chemical engineer
			Hanaford	Agronomist
			Shelski	Hydrogeologist
			Curtiss	Ad-hoc member from EPA

PROLOGUE

THE REALITY OF THE PRESENT: 2020

Can we allow our imaginations to envision futures that might be?

***

A hurricane was coming, and time was running out.

"Where did I put those notes from last week's symposium?" Mitch shouted, his mind racing like a steam engine ready to explode. *Damn*, he thought, flinging stacks of paper aside as he searched frantically through the piles that buried his desk—a deep black hole that devoured anything. "I get frustrated so easily when things don't go as planned!"

A familiar, reassuring voice from the other side of the room shouted over the howling wind outside: "Dr. Mitch, I forgot to tell you I put those symposium notes into the database for you last night." *Mitch's temper momentarily skyrocketed, but he calmed himself.*

"Thanks, Casey." Dr. Mitch thought, reassured: *My trusted assistant Casey! How could I ever get angry with him? He's such an important part of my research here at the Miami Center for Environmental Research (MCER)—always seems intuitively able to read my thoughts as partners working in such small confines tend to do.*

Anticipating his growing anxiety, Casey added, "They're in 'Climate Symposium Japan' in your documents folder."

Relieved, Mitch reminded himself: *See how invaluable he is to me! He keeps me sane in an insane world. Imagine, in a world that has only reluctantly accepted that climate change is really here to stay, we still have to fight for enough funding to get even this small space in the university. Still, I'm proud to be a part of the MCER recovery team, working in the internationally renowned Roberts Laboratory. But this old wood-framed Laboratory Annex, so close to the Atlantic Ocean, will surely go down in a Category 3 hurricane and we'd better hurry to clear the equipment and make sure all valuable data is saved.*

But Mitch wondered again, as he had many times before, exactly what his role was here. *How did I, Dr. Mitch Warner, get so involved in all this?* His drifting mind imagined that he was introducing himself. *You're probably wondering who I am. Call me Dr. Mitch, the mad scientist of climate research! Everyone else does. I've been told that I'm an iconic throwback to the 1960s with a flair for stubbornness, and yet still likable! I even have a 1960s 'hippie beard.' Am I a throwback in my spirit or is it just my choice of comfortable 1960's clothing?* He chuckled under his breath.

Mitch's mind suddenly cleared, snapping back to the sounds of windows shaking and billowing against the force of the approaching winds. The predicted Category 3 hurricane was quickly moving closer. Although amazed, as always, by the resilience and actual flexibility of glass during a storm, he understood that this building was still highly vulnerable: he and Casey had to move their valuable equipment to a safer location STAT. Then, momentarily embarrassed, Mitch remembered his carelessness in continually ignoring to update the backup of his hard drive.

"Casey, we'll probably need to wrap up earlier than four o'clock. The black cloud-front of the hurricane is already rearing its ugly head." "K. C." Klein, also known as Casey, looked out at a quavering window and seemed to shudder. He'd grown up on the vast Oklahoma plains and was never fond of rain blowing sideways.

"Wow, Dr. Mitch. Look at it out there! Shouldn't we get the climate-monitoring equipment away from the windows?" As he spoke, his hands were already clearing the desks and storing the sensitive instruments in a safer place.

"Good idea. Help me get the Remus ocean climatic monitoring vehicles loaded for transport."

"Stop worrying, Dr. Mitch. Hurricane Cecil isn't supposed to make landfall until late tonight. If the storm moves in earlier, campus security will call on my satellite phone and tell us to get to the shuttle bus in front of the North Concord Building. They'll make sure we're heading out before the hurricane winds reach the doorstep of the campus."

"I still don't like it, Casey," Dr. Mitch said, clearly becoming increasingly alarmed at the speed and intensity of the storm. "What does security know about hurricanes anyway? Besides, it seems like the tempest is already on our doorstep! This is surely a Category 4! Finish what you're doing and let's try to get to the shuttle location by three."

Casey looked out the window again. "It looks bad out there, Dr. Mitch."

"That's for sure, my friend." Sensing something wasn't right, Mitch raced over to the Emergency Radio Broadcast System, wondering why they hadn't received an updated alert.. Casey stopped what he was doing and they both listened to a broadcast crackling with static over the announcer's emphatic voice: "I repeat: —ricane Cecil exp—ted to hit six hours earl—er. Winds of one hund—d thi—ty to one— drer—fity miles pe—hour . . . predicted 20-foot storm surge. I r—peat seek shelter . . . —diately!" Stunned, both men were silent for a few seconds, simultaneously looking at each other and racing to the windows.

"Holy shit!" Mitch cried: "We've got to get out of here. Why weren't we warned?"

Casey looked at his satellite phone. "My God! The battery is dead!" He quickly plugged it into an external power source. For what seemed like an eternity, they both listened for missed messages.

One by one, the messages told the story. "ALERT! As of two p.m., all personnel are to evacuate immediately! Proceed to the shuttle bus. Hurricane Cecil is now a Category 4 and will reach landfall within the hour."

"Sometimes I hate it when I'm right," Mitch said quietly. Casey chuckled. The humor was a welcome release of tension, especially when they heard the next message. Someone was shouting: "Mitch! Casey! This is no joke. Where are you? Get here pronto! We can only wait ten more minutes! We're leaving at 2:30."

Mitch quickly scanned his watch: already two thirty. They looked at each other and went pale. He raised his arms in the air in his usual "let's go to Plan B" gesture while Casey ran to the window, frantically looking for the bus. In a slow Oklahoma drawl, Casey said, "It's gone." The satellite phone seemed eerily silent. Now the rain was coming down in sheets and intense lightning flashed as loud blasts of thunder bellowed through the storm.

Mitch cupped his hands to his mouth and shouted "Casey! There's no safety for us here! Time to move to the top floor of the North Concord building; we'll be further away from the storm surge there!" They both knew if they didn't leave immediately, their lives would be in great danger. The lights flickered and went dark just as they got to the front doors. Struggling to open the doors, the atmospheric vacuum pressure finally released them to the frightening scene outside. Rain pelted their faces with tiny razor-sharp shards. The deep blackness above them was impenetrable. The howling wind, filled with every kind of debris, raced past them. Tossed around like small insects in the wind, they fought for solid footing.

"Hold onto my arm! Two bodies are stronger than one!" Mitch yelled to Casey as he grabbed Casey's arm and they cautiously made their way to the North Concord building. "OOOOH!" Casey yelled in pain as a shard of flying glass slashed his arm. Just in time, they reached the entrance of the building. Reaching out with his other hand, Mitch pulled hard against the heavy glass doors and they broke into the relative safety of four walls and a roof.

The doors closed just as they heard more heavy debris smack and thud against the thick glass. Mitch turned to see Casey, standing still, catching his breath, and shouted, "Come on now! We have to go upstairs to the top floor!" Breathing heavily as they made their way to the top floor, they pushed their exhausted bodies against the walls for maximum safety in the dark stairwells of this newly designed hurricane-proof building. They were safe, for now. They finally reached the east side of the top floor, pulled together two chairs, and quickly sat down. All they could hear was their deep breathing. As the wind, rain, and debris unceasingly pounded the building, Mitch's mind filled with the realization that all the work he had researched so diligently and passionately could be gone in an instant! He sighed and came back to reality. "Casey, let me see your arm. There's a strange red fluid on your shirt." They both laughed; a much-needed release of tension.

Casey looked down at the gash on his right forearm, "You're a doctor, right? How are you at field dressing?"

Mitch smiled. "No worries, I failed that course in doctor school." Once again, they laughed out loud. Mitch stood up and went to the windows on the north side of the research room. "These white curtains will do just fine."

It took Mitch only a few minutes to clean and bandage Casey's arm. "Mighty fine wound-dressing, Doc." Casey looked up and smiled.

"Aw, it's nothing. I used to tie bandannas all the time in the 1960s." The thrashing of the wind and the rain was suddenly muted by an even deeper, more ominous sound: The hurricane surge had arrived. The two of them were silent as they watched the easterly 20-foot-plus wall of water envelop and destroy the Roberts Lab on the shore. Casey grumbled under his breath while Mitch wondered how many people were sucked out to sea as the wave receded back into the ocean. "Cat 4 like hell! This has got to be a 5!"

He's right, Mitch thought. *It is a Category 5. This Category 5 and those other two Category 4 hurricanes earlier this year—they're clear evidence of a significant increase in catastrophic meteorological occurrences all over the world. They weren't kidding twenty years ago when they predicted increases in major storms. I wonder if we could have avoided all of this if we had done more to mitigate our emissions back then. If only the fat cats in Washington had listened. Now I do feel like a throwback from the sixties, fighting for planetary justice.*

Suddenly the storm subsided; they must have been in its eye. Mitch sat in the chair facing the eastern side of the building, sadly watching the devastating consequences of Cecil. The high water surge and its torrent of spine-stiffening rain had passed. It was more of a thunderstorm now.

I wish I could have shown this to my colleagues twenty years ago—hell, the whole world: the devastating harm we have done to our planet. I must admit to feeling a lot less vigilant these days. Oh, snap out of it! Mitch groaned at his self-absorbed thoughts. "Well, Casey, there goes thirty years of research. I forgot to update my backup disc."

Casey smiled and looked at Mitch. "Ever hear of the Cloud?" Mitch's heart lightened He turned to Casey with anticipation. Casey kept looking out the window with a smile on his face and said, "Yeah, Dr. Mitch, I had a premonition that something like this would happen someday, so I've been collecting and scanning all our data to the Cloud."

Mitch threw up his hands and yelled, "YES! Objective B is in motion!"

"What was objective A?"

Mitch's eyes met Casey's. "Objective A was staying alive!"

Casey suddenly peered outside, signaling Mitch to look at what he was seeing, and shouted, "My God, they need help!"

"Where?" Mitch asked, rushing over to Casey as he pointed to three people on the campus walkway struggling in the debris and deep mud. They were certainly injured. "Come on, Casey, we've got to get them to safety, now!" Mitch grabbed Casey's sliced arm by mistake and heard a slight groan. "Oh, sorry, Casey."

"It's okay, I'm right behind you. Go, go, go!"

As they raced down the dark staircase two steps at a time, Mitch thought, "Adrenaline truly is an amazing tool when your body needs it."

Outside, the eye of the storm had just passed. The pouring rain and fierce winds intensified. Mitch struggled toward the injured people with Casey close behind. "Casey, we must have been in the eye of the storm, now we are back in the tempest—it all happened so fast!" Damaged trees were finally giving up their sandy stronghold—uprooted and thrown into the ever-increasing second onslaught of wind and rain. The two scientists reached the three students and guided them to safety just as a thick heavy branch from the campus' famed Honduran mahogany tree, planted nearly forty years ago at the entrance to the North Concord Building, fell on Mitch. His world went black.

CHAPTER 1

THE RED DOOR (#1): BUSINESS AS USUAL

What happens if, despite the scientific consensus that the world is warming too fast, we continue to stick our heads in the sand and conduct our Business as Usual? Enter into the first door and we will see what we will see.

***

In what seemed only a millisecond after the branch hit him, Mitch opened his eyes and found himself lying on the floor of a brightly lit vestibule. Dizzy and still dazed, he reached up to feel his head but, surprisingly, didn't feel a bump. He thought this strange. He touched his face and discovered that he was clean-shaven. This seemed strange—and mysterious. He squinted as his sight slowly came into focus and, as his senses began to return, became aware of how very quiet it was in this strange place. But most mysterious was the brightly lit hallway that faced him: it was lined with seven differently colored doors. He sat up in the vestibule, leaning against its brilliant white rear wall, and blinked repeatedly. Still perplexed, he looked down the hall and around the room, finally focusing again on the seven doors with a mixture of increasing amazement and curiosity. Each door was named by its color, a number, and a descriptor.

- The Red Door—Door #1: Business as Usual
- The Yellow Door—Door #2: The 20 Percent Solution
- The Green Door—Door #3: The 80 Percent Solution
- The Gray Door—Door #4: The Skeptics
- The Silver Door—Door #5: Geo-Engineering
- The Crimson Door—Door #6: The Volcano
- The Blue Door—Door #7: The Grassroots Solution

Is this a surreal dream, a hallucination, or am I simply losing my mind? If this is a dream, it sure seems real. He continued to stare at the seven doors, trying desperately to explain to himself what they might be or represent. *What the hell? I have a feeling my life is about to change.* He stood up and paused for only a moment. He took off his red beret, and dusted himself off. Still looking at the seven doors, he made a decision. *I'm a scientist, an explorer whose territory is information. Somehow, these doors are inviting me to discover something, maybe many things. It's an invitation I can't refuse. Door #1 is first—so here I go through the Red Door!*

Door #1: 2032

As he opened Door #1, Mitch sensed a rushing force pulling him into the swirling abyss of the doorway. "Ahhhh . . . !" Mitch felt his whole body being pulled and pushed inside the time/space tunnel. Just as quickly as it began, these forces stopped. Mitch was back in the North Concord building.

" . . . Ahhhh!" He stopped screaming when he felt solid ground under his feet and the sun shining full on his face. Casey was in the room, lacing his sneakers. Squinting from the light, Mitch peered out the window and saw students bustling about on the MCER campus. It all looked the same, yet somehow it was totally different. Mitch saw Casey getting up from his chair. "Casey, what the—did I miss the end of the hurricane? Hey wait a minute . . . the Roberts building!"

"You're thinking of Hurricane Cecil of 2020, aren't you?" said Casey gently, "Dr. Mitch, you got knocked out by a tree branch. The Roberts Lab was swept away by Cecil. You missed the end of the tempest by about twelve years." Mitch was speechless.

"Welcome to 2032!" Casey smiled his characteristically warm smile.

Mitch was disoriented; his intense hazel eyes darted this way and that way. He was dressed in his usual garb of blue jeans, green army coat, and trademark red beret. He inhaled deeply and spun toward Casey. "Was I in a coma? Was I hit that badly?" His thoughts were reeling.

"Dr. Mitch, I know about the Doors. You were in the time vestibule and we went through The Red Door together." Mitch stumbled slightly as he grabbed the edge of a nearby table for balance. "Everything is okay. I've been instructed by my superiors to guide you through this crucial time in your—well 'our'—future." Casey gently put his hand on Mitch's right shoulder. Mitch, ever analytical, thought, *what you see is usually what is.*

Amazingly, he accepted this news without surprise. He trusted Casey, even now. "Casey, I knew in my gut that somehow those doors were real. My only question is: why me? Well, okay, I have a <u>lot</u> of questions, starting with how the hell did we end up here?"

"You, Dr. Mitch, are the one key factor that may turn the tide in our planet's global existence during our upcoming journeys. And besides, you were chosen because we knew that you were the only one crazy enough to accept time travel!" They both laughed. "I can only tell you that although I am from your time period, I was also chosen to be your guide by conscientious people who live in a different dimension beyond us."

"Who are the 'we'? Who are the 'conscientious people'?"

"Easy, Dr. Mitch. We have to guide you through this journey gradually and it will all become clear to you in time."

Mitch glanced over Casey's thin frame and rather ragged clothing. "Casey, are you okay? You look a little rough around the edges."

Casey looked to the distance with a sad frown. "Well, first of all, we're in a deep recession. After the disappearance of Arctic ice and the calving of some of the Ross Ice Shelf in Antarctica, worldwide sea levels started rising much faster than predicted. All of our coastal cities were threatened with inundation. So in 2025, President Richard Mazikowski's administration acted with a ten trillion dollar Climate Emergency Relief Fund to subsidize the construction of dikes around many of our east coast cities within ten years. The funds also subsidized agro-technology to reverse the collapse of agriculture in the west. Also, they—"

"Whoa, slow down, Casey. What happened to agriculture?"

Casey shrugged. "Gone. Climate change turned our breadbasket into a dust bowl, insect pestilence still rages out of control, and the water table is in serious shape. Fracking caused incredible damage that further diminished water supplies; the ceaseless tide of eco-refugees from submerged coastlines led to eco-wars throughout the world; heat waves caused increases in heat mortalities—you name it. The Feds want to re-start agriculture by subsidizing the super-efficient hydroponic farms that had been successfully piloted, but there's still a lot of skepticism as to whether it can bring agriculture back to the U.S. And it's not just us. Agriculture has moved pole-ward, worldwide."

Mitch shook his head. "Damn it. All this happened so fast! Let me guess on the economy. The Climate Emergency Relief Fund caused massive deficits, then rampant inflation, then unemployment. The only industries keeping our economy from being in worse shape than it is are construction and engineering for the dikes and the hydroponic research initiative. Beyond that, few people are employed. Am I right?"

"Correct," said Casey. "You catch on fast, Doc. Our economy boomed in the late teens and early twenties with all of the increased natural gas and oil production, but without mitigation, we are now paying the price."

Mitch then realized he actually had memories from his twelve lost years. "Casey, it's odd that I know all of this. It's as if I really lived through all of those years, but . . ."

"Yes, we've learned that when you travel forward to a time and place different from your original plane of existence, most of the knowledge of what happened between the time you left and the time you arrived goes with you. Don't worry, when you're in this dimension, more memories will be triggered. At first, they might be vague, but when you jump into the scene as an active participant, you'll know everything that has happened since 2020. It's as if a time-space library is in your brain. Pretty cool, isn't it?"

Mitch's mind raced to catch up. "But, what about the campus?"

Casey pointed back to the window. "Look over there. Everything south of the main quad is now the campus wetland. Over there is the Williams Tower.

"It looks like a chimney with a pancake top."

"Yes! It was constructed in 2025 to replace the Roberts Lab. It has a revolving upper section for radar analysis of cloud structure, and deep photo atmospheric temperature sensors. It's raised, topographically, and is hurricane proof."

"Casey, it's not over yet, is it?"

He shook his head. "Well, first of all, the university is making plans to move twenty miles inland as the rising sea level will probably inundate us with the Atlantic Ocean by 2050."

Mitch responded: "It's a tragedy that we can't stop that here in Florida. Dikes are ineffective against our porous limestone bedrock. Seawater will just seep into the rocks under the dikes as it makes its way to the continental side."

Casey twirled his red braided hair. "You're right, of course."

The professor patted him on the shoulder. "What about you, Casey?"

Casey looked downcast. "I'm unemployed," he said. "It's a bummer."

Mitch looked pensive. "Sorry about that, my friend. All of this happened in the span of just twelve years. Who would have thought that the predictions from the 2014 IPCC[1] climate assessment report would *under*estimate the worst-case consequences of the Business as Usual approach? I deeply regret that a talented graduate student like you is deprived of a career." Mitch looked at his former graduate assistant. "How are you making ends meet?"

Casey shook his head with resignation. "I make some money doing odd jobs for the few who still have money, but, frankly, I'm having trouble paying my rent."

Mitch smiled. "Okay, here's the deal. If my memory is correct, I have an extra room at my house and I can cook pretty well. How about free room and board in exchange for being my assistant again? This is not a handout: I expect that I will need a lot of tech work since climate ecologists should be in high demand in this crisis. What do you say?"

Casey thrust his hand out and beamed. "Perfect! Thanks, Dr. Mitch!"

---

1 International Panel on Climate Change. See: http://ar5-syr.ipcc.ch/https://en.wikipedia.org/wiki/Intergovernmental_Panel_on_Climate_Change.

This time travel thing was intriguing indeed. "Help me out here, Casey. One minute I'm running through a Category 5 hurricane in 2020 and now I'm on campus in this dark age of 2032. What gives with those other six doors?"

Casey sat down and explained to Dr. Mitch. "I know it seems complicated, but through each door there are several time periods we can travel to. Behind Door #1, we have set specific time variables for 2032, 2040, and 2065. We'll go through the other Doors in due time, but for now it's Business as Usual. Follow me to 2040." He reached into his pocket and pulled out his phone, then typed in the time continuum coordinates for 2040 and tapped his screen. A small, cone-shaped object appeared at the top of the screen.

Mitch stepped back. "I'm not going to be hit in the head again, am I?"

Casey laughed. "No, Dr. Mitch, just follow me as we take a time-warp ride ahead to 2040."

"Wow, Casey, I guess I need to upgrade my phone!"

Casey looked at Dr. Mitch and smiled, tapped his phone again, and the brilliant white vestibule reappeared. Only this time, there was only a red door that said Door #1: 2040. Casey's voice seemed to echo in the chamber. "Watch. You and five other professors from around the country are participating in an inter-disciplinary debate about the state of the world's climate and its ecological consequences. The debate is moderated by Howard Hanaford of US-RBC News."

Still wary, Mitch said, "Whoa! How will I be able to participate in this debate after dropping in from 2020?"

"Well, you are not only a time traveler; you are part of the future." Casey was, indeed, a time travel guide. As he opened the door to year 2040, Mitch felt himself thrust forward into the new era.

Door #1: 2040

"Good Evening, I am Howard Hanaford. I want to thank you all for being here today to discuss what is being considered the hottest topic we are facing today: the 'global climate emergency.'" Hanaford introduced the panel.

"Let's start with you, Dr. Bertrand. What are your views on our current climate and its effect on society?"

Dr. Carl Bertrand, an ecological analyst from Yale University, began by offering a pessimistic view. "We are paying the price for years of sticking our heads in the sand. Our Business as Usual practice of emitting heat-trapping gases into the atmosphere pushed us past the threshold of irreversible climate change and all we have left to do is adapt. Yet, if Congress adopts Senator Christof's Climate Recovery Plan, we might –" Bertrand was interrupted by Calvin Curtiss, an economic policy analyst from Kansas

State, "Oh come on, Carl, we've had a few bad years of farming and are now adjusting to the new hydroponics technology. This will revolutionize agriculture and bring the breadbasket back to the U.S. As for sea-level rise, we've been hearing these gloom and doom predictions for years and, except for the coastlines, I don't see much water on our doorsteps yet."

Bertrand lashed back. "One picture tells a thousand words, Curtiss!" He slapped in a holographic "smart slide" that projected in the center of the room, waved his hand up and down, then side to side. A slide appeared in the air in front of the panelists. Mitch was intuitively aware of this technology, although it hadn't existed in the world of 2020. With the wave of a palm, the data disk in someone's hand would activate an image, much like old-fashioned PowerPoint shows, but instead of a screen, something like a hologram appeared in thin air. The image projected in the air.

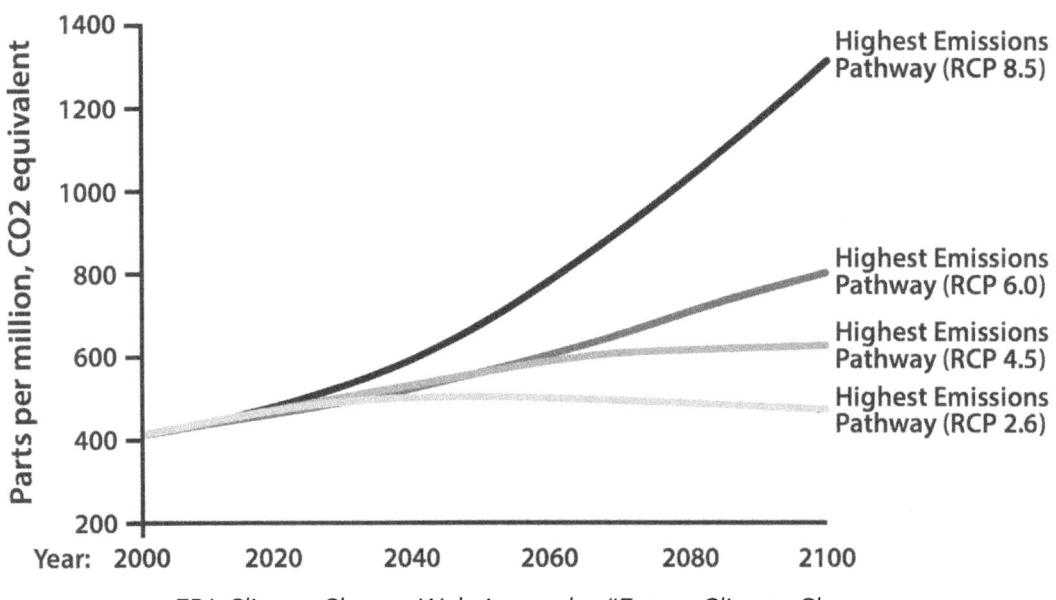

Projected Atmospheric Greenhouse Gas Concentrations

EPA Climate Change Website, under "Future Climate Change; illustration entitled Projected Greenhouse as concentration[2]

"How can you have any doubt that this data is real?" Bertrand demanded. "In the year 2000, a study was done projecting what the concentration of Greenhouse gases would be under different scenarios. For goodness' sake, you can see that the heat-trapping gases predicted for the RCP 8.5 scenario, representing Business as Usual, actually underestimated the concentration that we have today. Now look at how these

---

[2]

greenhouse gases are projected to skyrocket in the future, as shown by diverging lines on the right side of the graph. These alternative futures represent how high these levels would be under different scenarios of how we address the problem in the next 100 years. For God sakes, the only scenario where the concentration of the gases remains constant is if we cut our emissions by almost 100%! The study behind this graph was done in the early part of our century. We had choices back then and we chose Business as Usual. Now we're on the most sharply rising part of the curve." Bertrand was speaking rapidly, his face turning scarlet with visible rage. He took his Smart Slide Master and projected his second slide.

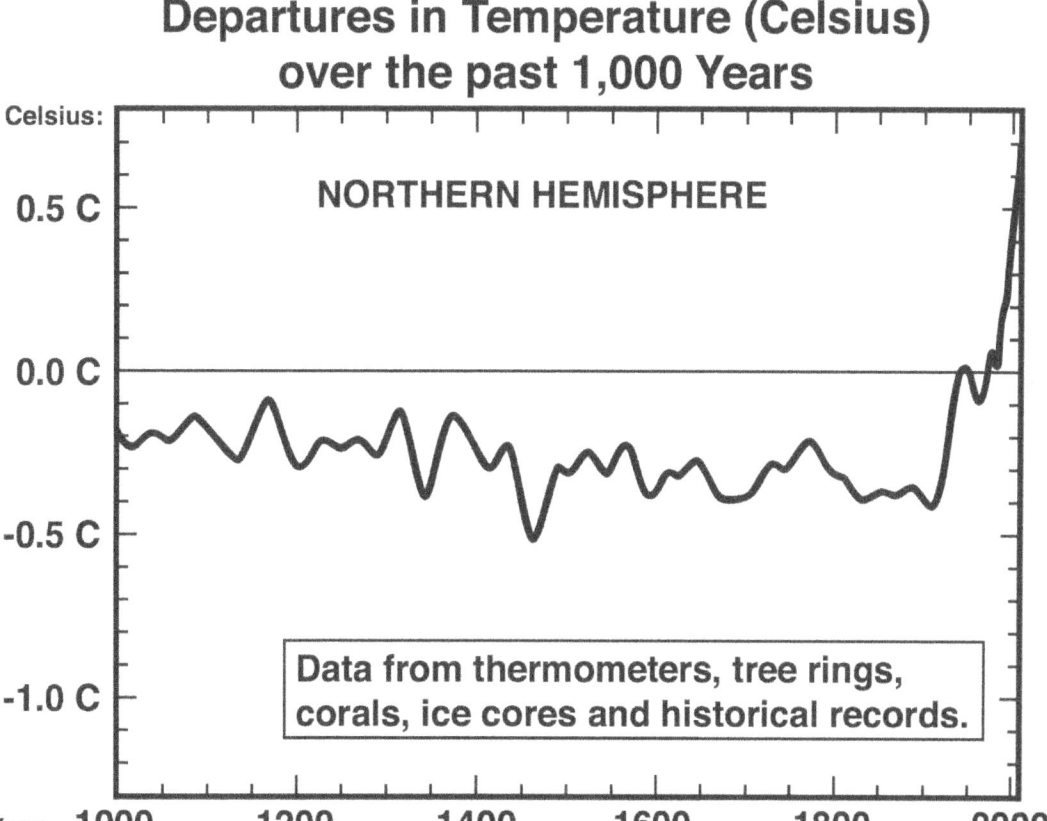

From International Panel on Climate Change, *Third Assessment Report: The Scientific Basis*. Illustration titled "Variations of the Earth's Surface Temperatures"[3]

"Look at how the temperature rises exponentially since the nineteenth century and it does so in tandem with the rise of CO_2, for the love of Mike! Look at the temperatures

---

[3] https://www3.epa.gov/climatechange/science/future.html

since the year 1,000 and how they skyrocketed during the industrial revolution in 1800.[4] Although this graph is originally from a 1999 study, the trajectory since the 'hockey stick' trend was first elucidated has continued onward and upward. The same trend as the gases! On a more recent scale of time, look at how our temperatures have risen since 1880s! The heat-trapping gases skyrocket and the temperature follows suit. Don't tell me there's no smoking gun here!" he shouted in exasperation.

Hanaford broke in. "That's enough!" he shouted. "We need to have order." Silence filled the room as Hanaford let his strong admonition take effect. "Let's go easy, everyone. Dr. Mitch, as a renowned climate ecologist, would you explain the implications of Dr. Bertrand's slides? Please state this in simple terms for those in the audience who do not have a science background."

Mitch stepped up and asked Dr. Bertrand to project the second slide again. "We have data going back 1,000 years in this case, but they go back hundreds of thousands of years in other studies. The shape of the steep rise in temperature since the 1800s is known as the 'hockey stick curve'." He explained that CO_2 and other heat-trapping gases caused the climate to warm. As someone on the panel tried to interject, Mitch raised his hand in a powerful gesture that amazed even himself, and squelched the objection before it could be made. He continued to explain that there are natural variations in the atmosphere that cause these heat-trapping gases and temperatures to fluctuate as he pointed toward the flat part of the curve. "This is the handle of the hockey stick. It illuminates the world's natural thermostat." He further explained that the steep rise in temperature at the right end of the graph is dramatic. "And this, folks, illustrates that the world's thermostat is going crazy. The temperature rising from the handle of the hockey stick occurs in lockstep with the rise in heat-trapping gases. This is the smoking gun Dr. Bertrand refers to. As greenhouse gases accumulate, temperatures rise in response. And, as we know, temperatures are rising straight to the sky."

Bertrand jumped in, projecting his first graph of projected increases of greenhouse gases. "These are projections of how greenhouse gases will rise in this century. The two lower curves represent 100% emission cuts and 80% reductions respectively. If we could achieve these cuts, the CO_2 in the atmosphere will start to stabilize. The very top line is Business as Usual, where we are not reducing emissions, but increasing them. The Business as Usual curve will cause our average world temperature to exceed a two degrees Celsius increase in world average temperatures. Mainstream climate scientists consider this to be the threshold where environmental and societal impacts will become irreversible, the point of no return.[5] Now as for the ecological implications . . . "

---

4 Mann, M. *The Hockey Stick and the Climate Wars, Dispatches from the Front Lines.* Columbia University Press: 2012.
5 See http://www.wri.org/ipcc-infographics

Those on the panel who disagreed could not contain themselves any longer. Bertrand was interrupted by Curtiss. "Okay, Bertrand, who took the temperature measurements from the year 1000 to 1800? The Vikings? Medieval bishops? Christopher Columbus?"

Mitch's anger and growing impatience expressed itself in his body language: his cheeks twitched and his face shaded into beet red. The room went silent as everyone waited for his explosive response. "Proxies!" he shouted. "Don't be a moron. You know we've been measuring ice core samples, corals, and tree rings since before you were born! We can date them and infer the amount of the CO_2 in the atmosphere of those earlier times. These proxies tell a true story about past climate."[6]

Curtiss responded sarcastically, "Yes. It's a story all right. It's pure fiction and your data is from a witch doctor." Several of the skeptics chuckled, feeling they had scored a victory. It was at this point that the heated debate escalated out of control.

Hanaford pounded his gavel sharply three times, finally getting everyone's attention. "Last warning! If we have another angry outburst from any of you, I will shut down this debate. Now, let's hear from Dr. Che-Ling, an environmental policy analyst from the University of Buffalo."

"Thank you, Mr. Hanaford. Well, Carl's perspective of passing a threshold is undeniable. The 'fixes' that are proposed are, at best, merely Band-Aids and will cost trillions more than we've already spent. He's right. We'll need to adapt more than mitigate. Mitch explained the historical trends of temperatures and gases very eloquently. Thank you, Mitch."

Hanaford addressed the next question to the entire panel. "Some experts say that we no longer have any realistic way of cooling down our climate. Do any of you think otherwise?"

Curtiss jumped into the argument again. "How do we know that we have a long-term crisis as opposed to simply a period of unsettled climate that will stabilize over time? There have been periods when the earth was inundated by floods, decimated by fires, parched from drought—look at the Dust Bowl of the 1930s—so perhaps our earth is merely experiencing a temporary period of instability. Where is the proof that it is something other than that?"

Dr. Mitch's face reddened again as he broke in, this time keeping his voice lower, determined. "How can you possibly think that way in light of all of the evidence? It is true that up until the early part of this century, we were basing our warming predictions on theory without the weight of evident consequences to back it up, but the first big clue that these consequences were rolling in was the melting of the Arctic ice. Remember this?" Mitch put a disk in his palm, waved his hand up and down, then side to side. His slide appeared in the air above them.

---

6 See https://www.ncdc.noaa.gov/news/what-are-proxy-data

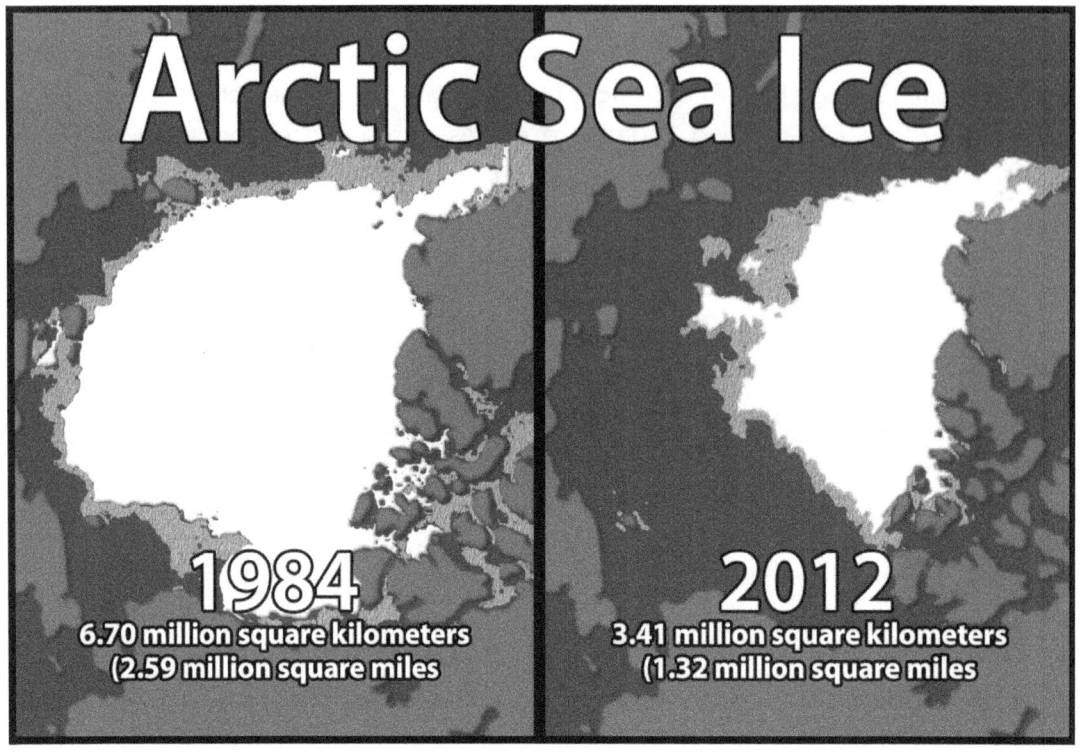

From NASA Website "NASA Finds Thickest Part of Arctic Ice Cap Melting Faster; Illustration entitled "Multi-year Arctic Sea Ice"[7]

Mitch explained: "Arctic ice in 1984 covered most of the Arctic Sea and was projected to melt slowly. By 2012 it was more than half gone. Today, of course, there is no longer any ice in the Arctic in the summer. And, not to be forgotten, Antarctica has about seven million cubic miles of ice that is melting at an alarming rate as well."

The room was silent for a moment, seeming as if he had made his point. Quickly continuing, Mitch almost shouted, "That's long-term, man! The sea ice has been there for millions of years. It hasn't melted during the entire period of human occupancy of the planet, until now. We are seeing conditions that have normally evolved from millions of years condensed into a 200-year period. Add to that all the heat waves, droughts, wildfires, increased frequency of severe storms, and the rising seas . . . well, I'd say that we have a slam dunk. It is our human induced warming that has triggered these events!"

Curtiss sneered. "Are we really warming? You whackadoodles keep telling us that we have warmed a measly one degree since the 1800s and that when we rise another degree that the sky will be falling. For God sakes, when I start out with a sweater on a

---

[7] See http://www.examiner.com/article/nasa-releases-imagery-comparing-record-low-sea-ice-to-average-levels

cool morning and it warms two degrees, I will hardly notice and the sweater stays on. If it warms by ten degrees, I may need to take the sweater off, but I will still be standing."

Bertrand kicked Mitch under the table when he saw his beet red face twitching again. This was enough to keep Mitch from an impulsive outbreak and he then refuted this misstatement himself: "Curtiss, Curtiss! You are confusing weather with climate. You can't judge what has happened over recent years and apply our current weather to what may happen many years in the future. Rather than try to explain this in detail,[8] look at what is happening in the world around you over the past forty years. We lost a good portion of Bangladesh to sea level rise, sea ice in the Arctic is gone, and the long term drought throughout the world is moving the agricultural belt of the world toward the poles. These are all the true indicators of long term climate change that normally take millennia to occur. The industrial revolution and its aftermath have caused these indicators to occur in two centuries."

Dr. David Marcus, a geo-engineer and physicist from UCLA, spoke up next. "No one has mentioned the geo-engineering solutions. If we launch a fleet of large reflectors into orbit, they will reduce the incoming heat from the sun and cool our temperatures."

Mitch passionately refuted Marcus. "Not that again! We've screwed up the earth's thermostat already. Now you want to mess with the sun? We cannot predict with any accuracy the negative impacts of this irresponsible proposal!"

Dr. Tom Harley, an economist from Northwestern University, joined the discussion. "There you go again. Environmentalists are always talking doom and gloom and can never open their minds to realistic solutions. Look at the costs. If the Senate adopts Christof's proposed Climate Recovery package, it will cost us another twenty trillion dollars. Subsidies for windmills, solar farms, aid to nations that are submerging into the sea, agriculture adaptation, more infrastructure projects, and all of that baloney about mitigating with carbon taxes. How can we sustain all of that spending? Look at what the four trillion for dikes and hydroponic engineering have already done to the economy. On the other hand, if we adopt the geo-engineering proposal, it will only cost us one trillion. This is cheap insurance in case you guys are right in your predictions."

The debate went on for another hour. Afterward, Casey and Dr. Mitch walked outside along a side street in the city where the debate had been held. Wherever Mitch looked there were the clear signs of a city in decay: signs that local, state, and federal funding

---

8 Climate takes into effect long term changes in weather and is taken at different locations across the Earth. If you use a thermometer to measure weather where you are, it is an instantaneous reading – what you see (and feel) is what you get. But you cannot use the same thermometer to measure climate, because you are taking the reading at an insignificant moment of time and space. Measuring climate requires an integration of multiple temperature readings across our planet over time frames ranging from forty years to millions of years. Your sense of what the temperature is right here right now doesn't tell you anything about climate.

for repairs and the much-needed upkeep of citywide streets, buildings and infrastructure were drying up or already gone. Casey spoke to his dejected mentor, "You made your point, Doc, but frankly you were outnumbered. Most people simply can't accept a multi-trillion-dollar expenditure for mitigation. Even the lesser cost of geo-engineering would be difficult to get through Congress."

Neither man spoke again until Casey finally broke the silence. "Dr. Mitch, follow me to the era of Door #1 in the year 2065. I have to warn you, the situation gets worse." He pointed toward a private clearing, pulled out his phone, and accessed an encrypted code. Mitch and Casey were back in the vestibule. Mitch opened the Door to 2065 and they were suddenly sucked into the abysmal swirl once again.

Door #1: 2065

They landed in the dilapidated, stripped Concord Building of the MCER Campus, aged from years of abandonment and vandalism. Looking out on the eerie scene of the campus in 2065 was like seeing a doomsday scenario come to life. The campus was submerged under shallow water, the sun glistening in triumph on its surface as if nature was finally reclaiming itself as ruler over humankind's stupidity. Mitch was spooked. Casey saw his confusion and explained, "As you know, up until ten years ago, MCER was an idyllic center of learning with fountains and pathways and the energetic bustle of students. The MCER campus is now abandoned; submerged under six to nine inches of water. This is not because of a recent storm. This old campus is now a marshland. Check out the salt marsh vegetation that's emerging from the water. This is all due to the rising sea that has submerged Florida's entire coast."

Mitch was stunned. "Who would have thought that Hurricane Cecil of 2020 would be so accurate a precursor of later years?"

"Yes, Miami has been in tough shape since the climate emergency began in earnest in 2040. Although dikes and sea gates were built to keep other coastal cities from going under water, Florida's coastline could not be protected because of its limestone bedrock. A costly federal and state recovery project began relocating the coastal residents of Florida. The emergency required national infrastructure projects, which created a massive federal debt and a fifteen-year period of economic turmoil. However, saving lives and evacuating eco-refugees from coastlines took precedence over a depressed economy. The MCER campus is now relocated twenty miles inland." Mitch wondered where the main campus would relocate to in another twenty-five years, when the sea inundation moved further inland.

The facts were overwhelming. Casey continued: "Beyond Florida, the U.S. is suffering an exodus of agriculture to Canada since the Southwest and Midwest have become even more of a desert wasteland. The more frequent and prolonged heat waves in cities are causing increased heat-related morbidity and higher mortality among the chronically ill, elderly, and infants. The demand for electricity is now far greater than the infrastructure can handle. Blackouts and brownouts are weekly occurrences. Smog has increased due to the higher number of extreme heat waves in summer, which has caused a spike in respiratory diseases. The insect population is overwhelming in the warmer, more humid areas. The U.S., however, fared somewhat better than many other areas. Hundreds of thousands have perished all over the world. Coastal cities, especially in the third world, are lost to the oceans. Massive migrations of refugees are triggering 'eco-wars' in these locations. This 'climate emergency era' is wrecking havoc throughout the planet along with other destructive ecological events. The world is close to chaos, suffering ongoing economic depression and the unraveling of much of the social order. Wars are occurring over scarce resources of food, water, fuel, and the remaining productive landscapes." Casey paused.

"How could we have done this to ourselves?" Mitch asked sadly.

Casey responded with a hint of relief and optimism. "Well, let's look at a different scenario of how things could turn out." Once again, he pulled out his phone as they jumped into the vestibule again. "Okay, Dr. Mitch, now for the next door."

CHAPTER 2

THE YELLOW DOOR (#2): THE 20 PERCENT SOLUTION

What might happen if we go part of the way and reduce our greenhouse gas emissions by 20%?[9]

***

Door #2: 2034

I'm getting to like this tingling feeling of time travel. Before Mitch could finish his thought, he was propelled onto a beautifully lush green field. Suddenly *POP!* A projectile smacked into his forehead. "What the hell—am I hit? Is this the end?"

He saw Casey running toward him laughing and shouting, "It's better if you use the racket on the tennis ball rather than your head! Welcome to 2034 of Door #2!" As Casey reached him and bent to pick up the tennis ball, Mitch couldn't help but notice that he somehow looked better than he had in their last encounter. Pondering how the world had fared with a new 20 Percent Solution scenario, he asked, "Casey, are we still in the eco-depression scenario again?"

Casey chuckled. "You're probably done with this tennis game after that sting in your forehead. Hey, it was my game point anyway! Let's go for a walk and I'll tell you about it. I brought a news headline for you, come over here on the sideline." Walking over to benches that lined the tennis court, Mitch felt Casey's excitement as he picked up his news tablet and ran his hands over it. It lit up as he handed it to his professor. "Oh, I see your perplexed look. We did away with all printed words in 2028 and saved a bunch of trees." Casey smiled.

Mitch took the news tablet and started to read.

---

[9] Many scientists claim that if we reduce our emissions by 80 percent we will keep our average global temperature increase to less than 2 degrees Celsius (3.6 degrees Fahrenheit). These same scientists recognize an increase of 2 degrees Celsius as the red line we cannot cross. Others take a bleaker view and claim that we need to reduce carbon emissions by 100 percent—tantamount to eliminating fossil fuels immediately. What may happen if we go part of the way and reduce our greenhouse gas emissions by 20 percent?

Dateline: December 14, 2034

PRESIDENT-ELECT LANKES RECEIVES CLIMATE PLAN

President-elect Tom Lankes met with President Anthony Parino today to discuss the recently unveiled "Climate Stabilization Plan," dubbed the "Cooler Climate Plan." The Plan results from a study initiated last year by President Parino to identify climate-stabilization measures that are needed to avoid threats to society. It would require a broad array of industrial facilities to cut their annual emissions of heat-trapping gases. It would also require major municipalities to cut their transportation-related emissions through carpooling, increased gasoline taxes, and/or alternative transportation initiatives. Though the plan is controversial, President-elect Lankes is expected to enact it with executive authority.

The study confirmed the conclusions of the U.N. Environment Program's recent assertion that sea level rise threatening coastal cities is imminent; droughts in semi-arid regions of the world may become a permanent condition; and increased frequency in Category 4 and 5 hurricanes will continue. The study also concluded that these phenomena are undoubtedly due to climate change caused by human greenhouse gas emissions. The study recommended 80 percent cuts in carbon dioxide (CO_2) and other heat-trapping gases to mitigate the worst-case consequences. In a note of urgency, the study stated that these emission cuts need to be instituted in the next ten years for the world to avoid the predicted dire consequences. President Parino recommended 50 percent cuts in emissions. President-elect Lankes stated that his first action will be to mandate 20 percent cuts within five years as a starting point. Although the 20 percent target falls short of President Parino's 50 percent reduction goal (and the panel's 80 percent reduction recommendation), it can be implemented immediately under the existing authority of the Clean Air Act. President-elect Lankes will then take the lead on the global stage to urge worldwide emission cuts.

"Well, this is more like it. But is this real, Casey? Will we use this as the first baby step toward substantial emission reductions, or will we stall at 20 percent?"

Quite excitedly, Casey responded with optimism. "Well, yes, it does fall short of 80 percent emission cuts, but it's a good start."

"Why is that, Casey?"

"Well, Doc, by starting at the beginning of Lanke's administration and with the support of Parino's initiative from Congress, there's a very strong chance 20% reductions will be fully enacted by 2042."

"But what about the rest of the world?"

Casey acknowledged his concerns. "Most of the developed world has recognized the emergency and is slowly waking up to action; Great Britain's 15 percent reduction plan is a good example. Most of the developed nations will likely follow suit, pending our successful implementation of the plan. In Africa, tropical diseases resulting from new vectors are rampant, and heat waves are killing tens of thousands of people every year from heat strokes and dehydration. In South America there is a massive wave of migration toward the south of the continent where conditions are less harsh. In Asia, there have been increases in the number of major typhoons, tropical diseases, flooding, and desertification. Agriculture worldwide is migrating toward the poles, causing massive economic disruptions and eco-wars. The final straw was the loss of the first three Oceanic nations[10] (and southern Bangladesh) that are now under water. The U.N. is poised to use the U.S. model of 20 percent reductions and consensus is building to update the 2016 worldwide treaty with strict enforcement measures. Unfortunately, it took the onset of this worldwide climate emergency to wake us up."

Mitch was silent for a moment.

"Tell you what, Dr. Mitch, let's go forward four years to a college symposium that is discussing the current status of climate change."

Mitch joked, "Another road trip!"

Casey reached into his pocket and typed in the date, time, latitude, longitude, and elevation.

"Casey, you can set it for different areas also?" Mitch asked, fascinated by the technology.

Casey continued setting his Q-Time phone as he explained, "Time and space are both interchangeable. So as long as I don't get the height wrong. . . ."

"What! The height could be wrong? You could put us in a free space over a building's edge?"

Casey chuckled. "Don't worry, it knows where to put us and even has a failsafe. See the red areas on the screen? Those are no-no areas. Ah, there we go, in the green, right in the ballroom of the Waldorf Astoria in New York City." Casey touched the screen and once again the small, cone-shaped object appeared at the top. As they spiraled into the hallway and walked toward 2044, Casey said, "By the way, Dr. Mitch, you are one of the panel speakers."

"What do you—" Mitch's sentence was cut short by the swirling yellow time-space fog.

---

10   The IPCC has identified 11 island nations that may disappear due to sea level rise. See http://www.ipcc.ch/ipccreports/tar/wg2/index.php?idp=671

The Waldorf Symposium: 2044

"—expect of me." He cleared his throat as he realized that he was at a table with four other professors actively involved in a heated discussion. No one noticed his lapse of conversational time-space etiquette. He quickly regained his newfound memory by recognizing Dr. Howard Hanaford, an Ohio State climatologist, as the chairperson of this event.

"Well, here we are in the second administration of President Lankes. The 20 Percent Cool Plan is fully implemented."

Mitch was startled as the audience started clapping exuberantly. His thoughts warmed as he heard this. *Wow. This is the first time I have ever witnessed so much positive support. So many times I've only been the messenger of doom and gloom to a disbelieving entourage of skeptics. Could this truly be the awakening of the human condition toward some semblance of enlightened existence?* As the crowd's clapping subsided, Mitch looked like he was enjoying this brave new world even more.

Dr. Hanaford continued. "Thank you. It has been a long road, full of sacrifices. As you know, it was predicted in 2015 that the world population would grow to 11 billion by 2100. That will amount to 65 billion tons of CO_2 emitted per year worldwide.[11] That will vastly increase the world's temperature above the 2 degrees Celsius that is considered the safe threshold. Business as Usual cannot continue. Thankfully, due to progressive actions by many nations, the world's global carbon footprint has been greatly diminished. Nonetheless, many of you are asking why, despite these actions, there is still no evidence of a decline in the greenhouse gas concentrations in the atmosphere. Can you, as the panel, offer any suggestions?"

Dr. Carl Bertrand, an air pollution engineer, began the first round of replies. "Good evening. First of all, I want to thank this distinguished panel for being here, and especially thank you, the audience." Again, the audience clapped. Then, as they quieted, Dr. Bertrand began his substantive contribution. "Imagine, if you will, a balance between how much carbon we are emitting and the total pool of CO_2 in the atmosphere. Think of the greenhouse gasses in the atmosphere as a three-quarter-full bathtub and our ongoing emissions as the faucet continually running into the tub. The faucet will need to be turned off or the tub will overflow. The ratio of CO_2 emissions to CO_2 uptake by the absorbing 'sinks' of natural earth processes is still significantly out of balance. In some places, as in Bangladesh, we are already overflowing the natural uptake limit. We need to shut off the faucet now! Even then, it will take time to get to a normal balance."

"Aw, come on, Carl!"

Mitch knew this rather crabby voice. It belonged to Calvin Curtiss, policy analyst from the National Utilities Council. He was a chubby little man with his double chin clamped

---

11 Based on trend illustrated on the EPA website: http://www3.epa.gov/climatechange/ghgemissions/global.html.

too tightly by his necktie. Mitch chuckled for a moment as he watched his tie bounce up and down with every word. But he did not chuckle as he heard what Curtiss had to say next.

"This is just bunk! Our economy stagnated when the new regulations went into effect. Our GNP would be booming at close to 8 percent a year if industries, utilities, and consumers didn't have this carbon mitigation burden. And see, these preposterous draconian measures did not even lower the CO_2 levels at all."

Mitch felt a familiar twitch crossing his face and knew his cheeks were turning red. *Will this ever end with these contrarians*? He took a deep breath and controlled his temper by using sarcasm instead of screams. "Oh, Calvin, Calvin . . . you do know better. You've read the same report we all have. Heat-trapping gasses will decrease, but it's not a light switch. Our atmosphere will come back very slowly to a cooler equilibrium!"

Dr. Hanaford pounded his gavel three times. "Gentlemen, please! Our debate must be civil or we'll end up screaming at each other. Russ, what is your perspective on all this?"

Dr. Russ Harley, an economist, shared his perspective. "Well, it's true. The regulations have caused an increase in the cost of power, oil, gasoline, and industrial production. However, we are still growing at three to five percent annually despite the huge deficit. On the other hand, the new sector of industry—green technology—is dominating and may replace fossil fuel driven power, industrial production, conventional agriculture, and transportation, which are now in decline. Fossil fuels are clearly on their way out. Green technology will create many new jobs"

General Curtiss responded, "You've neglected to mention increased costs due to inflation, Russ. They've been skyrocketing since 2042."

Confidently, Dr. Harley answered. "Actually, the inflation rate increased from near 0 to 4 percent in 2042, 6 percent in 2043 and has gone down to 5 percent in 2044. In a growing economy, that is the norm."

Mitch respected Dr. Harley for his uncanny skill in making a short comment so effective, and appreciated his ability to realize that fluctuating moments in economic or climatic responses are just phases like the tides affected by the moon.

Now Hanaford directed a question to Mitch. "Dr. Warner, what are the ecological effects?"

Mitch leaned forward, his head down. He took a deep breath, slowly raised his head, and eyed the panel with a fierce look no one could ignore. "I have heard for so many years the arguments of men for the gain of the dollar at the expense of the loss of the planet. The real question is who will gain and who will lose? Profits are short term; although the future is not set, it is in motion toward another mass extinction. Extinctions have occurred in the past. However, this unprecedented change in climate in only 200 years—a mere blink in the history of our planet's climate history—has now come to haunt us. We have lost more species in the last twenty-five years than we have lost since human history was recorded. Those species that can adapt are desperately migrating pole-ward."

Pointing his finger at everyone on the panel, one by one, he continued: "And, there is *no* denying the disruptions to our ecosystems are in some cases catastrophic. The thirty-year Australian drought has caused a human and animal migratory nightmare. The disappearance of the Arctic Sea ice and the shedding of part of Antarctica's Ross Ice Shelf into the sea have catastrophically disrupted ecosystems, not to mention the effects it has had on the human condition. I don't need to remind you of all the human suffering we hear about every day around the globe. Yes, the 20 Percent Cool Plan is in effect, but it should have been 80 percent. We have crossed the threshold to inevitable, irreversible impacts. What a great legacy we will leave our children! Twenty percent is just not enough!" The audience started clapping, interrupted by only a few muffled boos. Mitch was deeply moved by the applause.

After the debate, Casey greeted him with a broad smile on his face. "Wow, Dr. Mitch. I never knew you could command an audience like that."

"I didn't either, Casey! I was just speaking with my passion for our planet. I only hope that someone out there is listening when it comes to our need to crank up emissions reductions to 80 percent." Casey took Mitch by the arm. "Come on. Let's see what else we could have done." In the vestibule once again, they now faced The Green Door. They walked through it and fell into a green-and-white swirl of time-space.

CHAPTER 3

THE GREEN DOOR (#3): THE 80 PERCENT SOLUTION

So what happens if we truly commit—embattle climate change with everything that we have in our arsenal?[12]

***

2044

As the green and white swirl dissipated, a beautiful, sunny winter scene enfolded Mitch and a cold winter wind hit his body. Shuddering, he closed his usual worn jacket and lifted up the collar for protection. *What now; where am I?* He looked around for landmarks and saw the Willis Tower in the distance. At that instant, he was smacked in the chest by a snowball, but this time he knew who had thrown it.

"Sorry, Dr. Mitch!" Casey and he laughed as they greeted each other.

"Casey! Every time I travel with you I get hit, smacked, or pummeled by something!" Casey laughed as he introduced two friends, who had snowballs in their gloved hands.

"Doc, meet Jake and Janey, my friends and workmates at Eco-Energy. They're twins." Mitch had been scanning the horizon as they approached. Off in the distance, he'd noticed a shimmer atop a number of buildings.

"Pleased to meet you, Dr. Mitch. We've heard so much about you. Welcome to Minneapolis!" The twins spoke in an unnerving unison.

"Yes, charmed to meet you, too, but what are those shimmers on the top of those buildings?" Jake looked at Janey and Casey, then responded, "Those are our new transparent wind turbines."

"Transparent?"

---

[12] Scientists are telling us that if we want to avoid stepping over the threshold and off the edge into a climate disaster, that we need to reduce our emissions of heat-trapping gases by 80–100 percent. Is that a pipe dream, or can we engage climate change with the same intensity that we used to land a man on the moon in ten years? We leave that question to the reader, as you journey down a path that leads to an Utopia.

"Well, yes." Janey looked at Casey, rather perplexed that Dr. Mitch hadn't heard of them. "One of the complaints in the public sector for the past twenty years was how obtrusive wind turbine blades were on the landscape. . ."

As in the script of a play, Jake continued the sentence, ". . . so we made them with an infrared transparent co-polymer that only birds and airplane pilots can see . . ."

Janey chimed in, ". . . but the blades are only visible to the naked human eye as a shimmer of light."

Mitch's teeth were chattering from the cold. "Well, two things. I'm very impressed and I am damn cold!" As they all rushed off to get him warmer winter clothing, Mitch asked, "What is Eco-Energy?"

Jake responded, "Eco-Energy is where we work. We're in the business of processing efficiencies of energy distribution for utilities. We are programmers for the household sector."

Janey chimed in, "We determine the household loading for a particular time of day and feed in energy from the new smart grid that was constructed four years ago."

Jake added, "Depending on supply and demand, we feed any combination of solar, hydroelectric, wind, or geothermal energy to meet the customers' needs."

Mitch smiled and nodded his approval. "It's about time!"

Once Mitch was properly outfitted, Jake and Janey bid their farewells and went back to work, leaving Mitch and Casey to explore the world of 2044, shimmering around them with a green hue. Despite the warmer clothing, Casey saw Dr. Mitch as a fish out of water in the cold, and a Florida fish no less. He put his arm around his shoulder protectively.

"Come on, Dr. Mitch, there's a coffee shop just around the corner where we can get really warm. I have so much to tell you!"

Mitch responded with a shiver. "I was hoping you were going to get us to the warmest spot possible." As they walked, Casey pointed out the superconducting transfer stations that were being built, the solar-powered streetlights, and numerous pedestrian walkways. It was dreamlike to Mitch, who was deeply impressed by the efficient beauty and absolute quiet of the public electric transportation practically floating—no, it *was* floating—above the street! Questions flew into his mind and Casey patiently answered them all.

"Superconductors, Doc. That was the ticket that revolutionized the transportation and smart grid. We were able to cut emissions and power usage significantly, but it still isn't enough."

As they reached the cafe, Mitch said, "I hope they still have coffee. It's the only fuel I ever needed." Casey laughed as Mitch rubbed his hand together, happy in the warmth of the cafe.

"In the early 2030s," Casey began explaining, "Greenland started disappearing as its ice sheets melted into the sea. Sea ice in the Arctic was gone by 2025. Then, in 2034, a portion of the Ross Ice Shelf in Antarctica calved into the ocean, shedding 25 percent of the continent's remaining ice. Things looked bad as sea levels began to rise, along with other consequences of a climate emergency. Then, President Julie Rother took office in 2036. She convened an Energy and Environment Task Force that recommended a course of action to prevent the realization of troubling predictions for the next fifteen years: inundation of coastal cities all over the world, increased frequency of severe hurricanes, drought, wildfires, and anything else that would create a worldwide crisis."

"What happened to fossil fuels?" A smile covered Mitch's face as he sipped his steaming cup of shade-grown Columbian Gold coffee.

"You're going to love this. Let's go to a video from 2033," said Casey, as he pulled out his portable nano drive with its expanding screen. "For once, we are going back in time, from 2044 to 2033. I have a video that tells it all. By the way, Dr. Mitch, you were in the President's Cabinet in 2033."

2033

The video started. They were in the American University auditorium. This was a world broadcast of President Rother and two members of her Cabinet meeting to discuss the climate change emergency. There was some skepticism about how scripted the discussion would be; having a Cabinet meeting in front of a live audience broadcast on the Unity-vison World Education program could be risky. Nevertheless, President Rother insisted there be transparency in the decision-making process relating to the crisis and sought viewer input on the decisions that were being made.

Applause greeted the president when she entered the stage; audience and panel members stood to honor her. As she settled into her seat, the auditorium became quiet. She looked out at the audience and to her advisors on stage and smiled.

"First, I want to acknowledge all the dedicated, hard work you have done. The time is right, not only for the American people, but for all the people of the world who have started taking action. We thank the governments of Europe, China, and the Third World Green Union for implementing the prohibition of new coal-powered plants. We appreciate the ingenuity of Japan in paving the way for 100 percent renewable energy and electric vehicle transportation, and China for its role in perfecting a cost-efficient carbon capture initiative technology for the remaining coal power plants. And we offer many thanks to the countless other contributors in the private and governmental sectors. Still, there is something missing: deep cuts in our carbon emissions, worldwide, that are enforceable. As a nation and as a planet, we need to move forward to meet this crisis."

Mitch felt tears welling in his eyes. He was aware that the world was finally turning toward a new fossil fuel-free globe. President Rother continued, "All the predictions of the first two decades of our century have materialized. The world paid lip service to many warnings, but did nothing to stop the increasing emissions and greenhouse gas accumulation in the atmosphere. Now my question is, how much more can we take? Dr. Carl Bertrand, our Transportation Secretary, has taken the lead in phasing out gasoline-powered vehicles. Carl, what is your perspective?"

Dr. Bertrand stood, cleared his throat and, spoke. "Madame President and esteemed colleagues, it is generally accepted by climatologists that there is a tolerable threshold in the level of heat-trapping gases in the atmosphere. Once it is crossed, it will trigger irreversible worst-case consequences. Some of my colleagues say that we have already passed that threshold. I think there remains a small window of opportunity, but we need to act fast to phase out all greenhouse gas emissions."

After another round of applause, the video zoomed back to President Rother. "At this time, I wish to acknowledge my Science Advisor, Dr. Mitch Warner."

Casey paused the video. "See, Doc, you were famous."

Mitch looked at Casey and grimaced. "I'm speechless!"

As he appeared in the video, the camera panned the huge room, as if in anticipation of something that would be unpleasant. President Rother asked: "Mitch, in your opinion, what will the worst-case effects be if we fail to control the level of heat-trapping gases in the atmosphere?"

"Thank you, Madame President." He looked down at his notes, took a deep breath, and responded. "If we continue Business as Usual, ecosystems will change drastically. Our own personal living conditions will be difficult and very different. Our report describes what we'll be facing in America, but this is a global phenomenon and the impact on other nations will be just as bad, if not worse. Geopolitical tensions will rise due to eco-refugees moving away from submerged human habitations; conflicts will increase over diminishing water, power, and food resources; floods and droughts will happen with greater frequency, followed by famines, starvation and the spread of infectious diseases. These and other pressures will destabilize peace and security across the globe. This was predicted in the Pentagon Report of 2013 on Future National Security Threats;[13] all are reiterated here. The future will not be a pretty picture, and we may see some, if not all, of these effects in the next ten years. That's why I recommend 80 percent emission cuts."

A restrained applause echoed through the auditorium. President Rother replied, "You know, Mitch, ten years ago I would have thought this just a gloom-and-doom scenario. But now, with a major part of Bangladesh under water, and the crash program to protect our coastal cities from inundation in effect, as well as all the other climate change indicators we recognize, you have the world's attention."

---

13   See http://www.scientificamerican.com/article/immediate-risk-to-national-security-posed-by-global-warming/.

Casey stopped the video. "This is huge, Doc! President Rother bit the bullet and submitted a bill to Congress mandating an 80 percent cut in emissions in ten years and prodded them into adopting the plan. She then formed a 'coalition of the willing': twenty-one nations that adopted a pact to reduce global emissions by 80 percent with very specific binding targets and mandates for each participant. The coalition announced that all its members would leave the UN unless the worldwide emission mandate was approved. The mandate was approved. Nearly all nations, including China and India, implemented the cuts. The mandate included massive aid from the developed nations to developing nations for industrialization with green energy rather than fossil fuels."

There was a brief silence in the cafe. Casey looked out the window. A light, fluffy snow began to dampen the sounds of the city. Casey looked at Mitch, deep in his own thoughts, and said, "Let's go for one last time travel in this dimension, all the way to 2152. Come, my friend." They swirled into a green fog to the last time and place into Door #3.

2152

When Mitch looked up, he and Casey were observing a classroom from a surreal perspective, as if they were hovering over the lecture hall. A sign above the female professor identified the lecturer: "Climate Dynamics, Dr. Leah-Ann Stuart." They seemed to be floating. It was an eerie yet exciting feeling. Mitch studied the professor and felt a sense of recognition: intuitively he felt she was his descendant. His mouth opened in proud surprise. Then he focused on what she was teaching.

"Over the past one hundred fifteen years, the U.S. has taken the lead by cutting emissions 80 percent after 2037. One example of our waking up to the climate change crisis was the oil and natural gas boom of the mid-twenty-first century. We could have just drilled and burned with no mitigation. Instead, we required tight methane and CO_2 capture technologies both at the drill sites and in the power production utilities. The rest of the world followed our lead and other nations followed suit, cutting their emissions by 80 to 95 percent. Additionally, a technological breakthrough occurred with carbon capture of ambient air[14]—giving us the ability to literally scrub out CO_2 from the atmosphere, thus lowering its concentration to pre-industrial levels. The world united in a massive mobilization to fight the impending climate disaster."

A student pushed a button, which lit up as he asked a question. "Wasn't one of the forefathers of climate restoration a relative of yours?"

---

14 See http://geo-engineering.blogspot.com/2008/10/removing-carbon-from-air-discovery.html.

There was a brief silence on the floor as Professor Stuart blushed. "Yes, the efforts of my ancestor, Dr. Mitch Warner, were paramount in helping the world see the possible devastation of a climate out of control."

Mitch and Casey exchanged smiles.

Dr. Stuart then moderated a discussion on the "what-if" scenarios that might have occurred if society had done nothing. She closed her class with a summary: "Yes, the economy took a hit, but the pain of the recession of the 2040s and early 2050s was more than outweighed by the progress being made to restore the climate. By 2060, the world had achieved 90 percent CO_2 emission cuts. By 2070, the rate of increase of average temperatures and the rate of sea level rise had slowed dramatically. There was hope that coastal cities would be spared without our having to spend trillions of dollars on more dikes and other adaptive measures. A side benefit of this astonishing turn-around in the climate was that the economy achieved a recovery comparable to our emergence from the Great Depression after World War II. The massive realignment of our industrial and commercial infrastructure to a green economy created millions of new jobs and began a revolution in the buying and selling of goods and services. The work created by setting a world goal of achieving 80 percent emissions cuts in ten years could be seen as analogous to the successful effort of the 1960s to put a man on the moon by the end of that decade.

"In addition, technology was developing at a phenomenal rate. Even though climatologists predicted that it would take centuries for the temperatures to settle down, the CO_2 levels were reduced—thanks to the emergence of ambient air CO_2 capture technologies[15] that were deployed worldwide to suck the CO_2 out of the ambient air and recover it in a water solution that was recycled as industrial solutions of carbonate products. In effect, the 100,000 carbon capture stations around the world reduced the CO_2 that had been emitted into the atmosphere since the industrial revolution. So, using our bathtub analogy, we have turned off the faucets that have been pouring new CO_2 into the atmosphere by eliminating fossil fuels. At the same time, we are reducing the 'standing water in the tub' – the greenhouse gas pool that lingers in the atmosphere. "By the turn of the twenty-second century, the frequency and intensity of droughts, wildfires, and disastrous hurricanes started to decline. Temperatures were actually decreasing. All of these measures prevented the 'double eco' (economic/environmental) disaster that society was heading into. Eventually, our climate settled down into a new steady state by adjusting to the phase-out of fossil fuels. Our world shifted to power from windmills, solar farms, hydro-electric, and other non-polluting sources. Our communities became sustainable and there were a lot more people bicycling around. We created an entirely new economy based on

---

15 This technology is conceptually feasible.   See https://www.technologyreview.com/s/531346/can-sucking-co2-out-of-the-atmosphere-really-work/

the new green technology industry. Students, we can thank my ancestor's generation for giving us and the world a brighter future."

After the class, Casey took Mitch to a back door in the basement of the science building. He encouraged Mitch to enter, smiled, and said, "Here, take this." Mitch opened his hand and Casey placed a small glowing green disc on it. Mitch looked at it for a moment and closed his hand tightly around the disc. Casey smiled again, his always warm and generous smile, and said, "So long, my friend and mentor. May you enjoy the brightest of futures!" He waved his hand, and before Mitch could say a word, he swirled away.

Back from the Future (to 2020 reality)

Mitch kept hearing a phrase repeated over and over again: "Casey is gone. Casey is gone." Awakening on a cot in the Concord Building of the Climate Research Center, he realized that he must have been shouting out loud.

An emergency medical technician (EMT) asked him, "Who is Casey?" Mitch looked up confused, wondering where he was, as the EMT continued. "Never mind that, Doc, you need to rest. We think you have a concussion." The wind was still swirling outside but greatly diminished. It was then that Mitch knew he was back in the reality of 2020, in the aftermath of Hurricane Cecil.

Wait, he thought, "*was that a dream, a premonition, or some kind of* Twilight Zone *episode? I feel disoriented, hanging in the balance of time warps. It probably was a dream, but . . .* Then he felt a disc-shaped object in his pocket.

Suddenly, as Mitch closed his eyes and rubbed his head, Casey appeared in a vision and spoke to him. "Who knows, Dr. Mitch? Whether you had a dream or a premonition or a trip to the Twilight Zone, the challenge is the same. Choices need to be made, and made now. It's our future that's hanging in the balance." More confused than ever, Mitch sat up and the small gently glowing green disc slipped out of his pocket onto the cot. Now Mitch smiled. He recognized the disc that Casey had given him. He remembered and decided to Let it Be. "Everything will sort itself out in time," he told himself.

Endnotes: Chapters 1–3

1. An 80 percent emission reduction plan is not fantasy. It is real. The State of California adopted The California Global Warming Solutions Act of 2006 as a blueprint for cutting their emissions by 80 percent[16] by 2050. Governor Edmund Brown has taken the first step with a goal of reducing emissions by 40 percent below 1990 levels by 2030. This is only a first step toward achieving the 80 percent reduction goal (to be achieved by 2050) established by the California Global Warming Solutions Act of 2006[17]
2. In 2015, New York has announced policies to reduce emissions statewide by 80 percent no later than 2050[18]

---

16 See http://www.arb.ca.gov/cc/cc.htm
17 See https://www.gov.ca.gov/news.php?id=18938.
18 See https://www.governor.ny.gov/news/governor-cuomo-joined-vice-president-gore-announces-new-actions-reduce-greenhouse-gas-emissions.

INTERLUDE: 2020

Mitch was inundated with flashbacks and truly shaken up. His thoughts were a muddle of ideas and, perhaps, truths—at least they felt like truths. *It just feels too real. Asleep in my own bed early in the morning and I'm hit by a torrent of flashbacks.* Exhausted, he checked the clock on his night table: 5:26 a.m. Last night, seemed like a dream. But the glowing green disc that Casey gave him sat silent next to the clock. *Hurricane Cecil had to have been real, hadn't it? But am I really back in 2020?* He asked himself, still unconvinced.

Sleeping was futile. Mitch glanced at his widescreen TV opposite his bed; it seemed archaic after the floating holographic screens he had seen in the other worlds. Frustrated by his own confusion, the scientist in him asserted itself: *Turn on the TV and get some real data on where I am.*

The television volume was loud, blurting out, "This is Jack Piatek of CSS News, Miami, reporting the April 23, 2020 morning news headlines. President Mazikowski yesterday announced . . ." It was indeed the present-day world of April 23, 2020, but there was no mention of the hurricane.

The news verified that he was back in the present. His thoughts rolled out again. *Let me get this all together. Maybe the doors were just dreams or visions from my subconscious. Maybe it was some fugue state brought about by too much work. If they were premonitions, I'm having a hard time dealing with it. Let's see, I don't have a class until later this afternoon, but to hell with it! I'll go to the office to distract myself from the weirdness of my experience, whatever it was.*

It was now 6:00 a.m.; someone was coming up the stairs. His wife Leah entered with breakfast on a tray. She asked how Mitch was feeling. He just shook his head and said: "Weird dream last night."

Leah looked at her husband with concern. "You were moaning and flailing and saying things that didn't make sense to me. Are you sure you're okay?"

He told her that he needed to get back to the university. Surprised, Leah asked why he was going in at such an early hour. "I'm overloaded with work." Leah shook her head in deepening confusion. "Remember to pick up Susannah from school at four o'clock."

Driven by an overwhelming sense of immediacy, Mitch ignored the breakfast and hastily put on his clothes and drove to the MCER campus. When he arrived, he was relieved to see the campus was still there. There were no hurricanes on the horizon. Everything seemed normal and familiar. His thoughts, however, kept returning to the vivid images in his dreams. He felt a chill run up his spine.

As he pulled into his parking spot at MCER, he stared at the sign with his name. Then, walking down the hallway to his office, he saw his secretary Lonnie already busy at her desk. He smiled. "Good morning, Lonnie." She looked up and smiled at him. Mitch was fond of teasing and flirting with her but never let it go further. Even though the attraction seemed to be mutual, he always reminded himself that going any further was a dead end to trouble. After all, he had a wonderful wife and family.

"Good morning, Dr. Mitch, you're early today. Your class isn't until two." Lonnie had worked with Mitch for several years; they were colleagues and friends. Mitch considered telling her about last night. *I might as well get this off my chest to someone I can trust.*

"I woke up very early this morning from a weird set of dreams that seemed to last all night and decided to come in early to get some work done."

"Want to tell me about your dreams," Lonnie asked, turning her chair toward him. "I can be your armchair analyst."

"Oh, you don't want to go there. You might become trapped in a cloud of curiously surreal images in the recesses of my mind. . . . " His voice trailed off.

Sensitive to his moods after having worked together for so long, Lonnie insisted gently, "Okay, Mitch, why don't you tell me about it? I'll give up my coffee break to be amused by what has obviously shaken you up."

Mitch felt embarrassed. "Maybe some other time. But thanks."

Lonnie gave him a serious look. "Are you okay? You do seem a bit rattled."

"I'm fine. Just need to immerse myself back into the work of my present-day world and it will pass, as all dreams do."

Focused on Lonnie, Mitch was unaware of the person behind him gently sipping his morning coffee and startled when he heard him speak. "So what were your dreams about, my friend?" Dr. Dave Cornelius, a professor of cell biology and a staunch disbeliever in the reality of climate change, entered the room.

Lonnie interrupted protectively. "Let it rest, Dr. Cornelius. I don't think he wants to talk about it."

Mitch quickly interrupted, "They were about climate change, let's leave it at that."

"Not *that* again!" Dr. Cornelius whined. "We're so saturated with that damn issue.

I don't know why you believe all that doom-and-gloom crap. Look, the broad populace does not want to accept or do anything about it. It's just too confusing and we're far from consensus. The science is weak." Dave babbled on, his resistance to the idea of climate change clear, and even though Mitch was used to hearing this close-minded rejection from him and other skeptics, such arrogance made him furious.

Mitch had always let Dave's skeptical remarks slide by attributing them to the unwillingness of most people to see the real truth. Today, however, his response and patience were different. He seethed inside. He invited Dave to join him in the faculty conference room to discuss this issue further.

Lonnie looked at Mitch with concern. *He isn't himself today*, she thought. *Mitch never takes on that bag of wind; he just ignores him. But today, Dr. Mitch has fire in his eyes.*

Mitch and Dave left the office and headed toward the faculty room. Dave argued on, chuckling as they sat down at the oversized conference table. Mitch poured himself a cup of coffee and refilled Dave's cup. "Careful on the caffeine, Dave, this discussion may generate enough firing of your neurons."

"Yeah sure, Doc." Dave rolled his eyes. Then they got into it.

As they settled in, Dave opened the dialogue. "Okay Mitch, what's this all about?"

"Well Dave, for years I've just put up with your negative comments on the prospect of whether our climate is changing, but I've always disagreed with your viewpoint. So let's have it out."

"Bullshit, Mitch. I know where you stand. You and your far-out fringe environmental activists are all wrong on this. I certainly hold open the possibility that the climate *may* be changing, but the evidence is weak."

Mitch went to the whiteboard and drew the Keeling Curve. "Okay, Dave. You may know this already, but how can you refute this curve? Since 1952, the concentration of CO_2 has been increasing steadily. We have known since the 1800s that there is a link between increasing CO_2 in the atmosphere and increasing temperature. This graph shows the steady rise and other data demonstrate that this increase in CO_2 predates the measuring period – it has been ongoing since the industrial revolution."

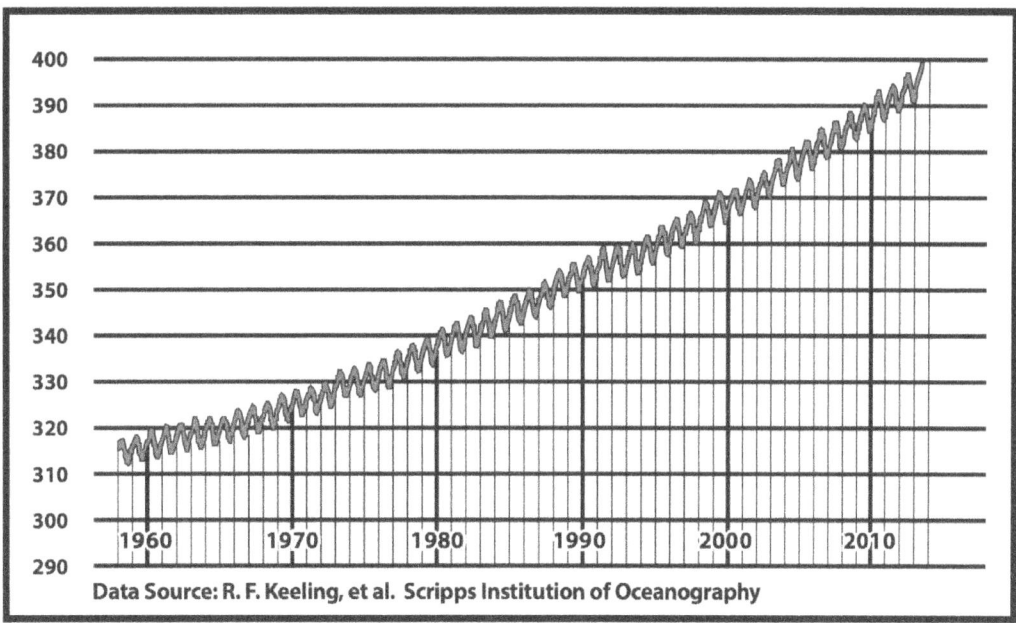

From "Climate Central" Website[19] Keeling Curve at opening of the webpage.

"Well, of course CO_2 is rising there. The detectors measuring the rise are on top of a volcano—Mauna Loa in Hawaii—where CO_2 has been leaking from the volcano itself into the atmosphere for centuries. It's all natural, Mitch."

"Not so, Dave. We still see the rise of CO_2 on *top* of the baseline volcanic emissions of 1952. Any background emissions from the Mauna Loa volcano will be factored into the measurements. Other monitoring stations around the world are also picking up the same trends and the same levels of CO_2. Virtually all measurements of climate change are backed up by collaborative studies. The scientific method—remember what that is, Dave?"

Dave's face was becoming more and more red.

"If you're so certain about the scientific method, you should back up your claim that other stations have corroborated the Hawaiian volcano readings. The devil is in the details; you should know that, Mitch! How do I know whether the other stations are located on volcanoes or other sources of CO_2?"

Mitch's shoulders were tensing so much, his shoulders and ears appeared to be attached.

"You're missing the point! If—"

---

19 http://www.climatecentral.org/gallery/graphics/keeling_curve.

Dave was now standing.

"Even if there is an increase in CO_2, what's the big deal? Human emissions pale compared to naturally occurring water vapor in the atmosphere and other natural sources such as biologic respiration, worldwide."

"Damn it, Dave." Mitch slammed his hand on the table. "You're jumping from topic to topic. Natural sources do indeed contribute to the natural thermostat of the planet, but our thermostat has evolved over billions of years. The sources of change wax and wane over thousands to millions of years. Today's unprecedented rise in greenhouse gases has occurred in a mere two hundred years. That is bound to flip our global thermostat in ways we cannot imagine."

"Well, Mr. Know-It-All, you just said it. 'Ways we cannot imagine.' We don't know enough about the climate system to predict what effects there will be, if there even are negative effects."

"There you go again, cherry picking statements out of context. We do know about what the effects will be and the world we hand to our children will not be a pretty one."

"THERE YOU GO AGAIN! Global warming is a theory, man. It's up to you guys to prove it."

Mitch struggled to contain himself.

"OPEN YOUR EYES! All the mainstream scientists are in agreement that CO_2 emissions are skyrocketing. Temperatures are rising and sea level is threatening to inundate our coastlines. Human health, global conflicts and ecosystem changes are already happening. We just don't know specifically how they will play out."

"HAVE YOU LISTENED TO YOURSELF? You say 'we don't know specifically how this will play out.' You just contradicted yourself again. I think that you are just parroting what your socialist friends are telling you."

"GO TO HELL, DAVE!"

Mitch stormed out of the room. He took a few deep breaths and headed back to his office. Lonnie looked at him in alarm. "What happened, Mitch? Your face is beet red and you're totally riled up."

Mitch took a few more deep breaths and shook his head. "Cornelius is a jerk. I have some better words to describe him, but I won't use them in front of a lady. Where is Casey?"

Lonnie looked bewildered. "Casey? Who is Casey?"

Mitch looked equally bewildered. "Casey, my graduate student. I need him. What is wrong with you, Lonnie?"

She looked very concerned. "I'm sorry, Dr. Mitch, there is no one in our department by that name."

Mitch became disoriented and nearly toppled over in a wave of dizziness. Lonnie ran over to help stabilize him. "Mitch, do you want me to call the Health Care Center? Or, I can walk you over there."

He took another deep breath and finally regained his composure. "I'm sorry, my friend. I'm still reeling from those dreams last night and Cornelius's pompousness just threw me for a loop. I'm okay now."

But Mitch was not okay. He went into his private office and sat down. Lonnie followed closely behind. *Whoa, what is happening to me? Was Casey just part of that dream? Dream, hell! That was a fugue state and I need to figure out what to do. I've changed; I have to do something about what will inevitably be a difficult world in our future. No more letting the skeptics of the world go unanswered. I need to shift priorities in my life. What can I do to get the message of my Twilight Zone travels out to the masses*? Mitch fiddled nervously with the change in his pocket and felt the small disc. Gingerly, he pulled it out and placed it on the desk in front of Lonnie. "Here's the proof! Casey passed this along to me as I was exiting the fugue state."

Now Lonnie looked very alarmed. "Please, Mitch, let me take you to the Campus Health Care Center."

Mitch sighed. "I've got to get a hold of myself."

And then, in the days, weeks and months following the night of his dreams, as if a button had been pressed, and a switch turned on, Dr. Mitch Warner found himself pursuing a different path, embracing a new life of activism. He allied himself with organizations that were rallying scientists and society to wake up and change course. He went on speaking tours across the country, and spoke before the Society of Climate Scientists and the EPA, where to both audiences, he presented his position in an unmistakably urgent tone. He urged them to convince business, government, and reactionary contrarians to change our current course of Business as Usual, warning that if action wasn't taken, they would all become merely passive observers as the inevitable and devastating consequences roll over us. In one demonstration he led in downtown Miami, he was arrested for getting into a fistfight with a bystander who was taunting climate change demonstrators.

Mitch's personal crusade took its toll on his personal life. His relationship with his wife, Leah, and his daughter, Susannah, grew increasingly strained. He became argumentative, easily irritated, and increasingly impatient with everyone: family, friends, and colleagues who were skeptics. Leah, Susannah, his remaining friends, brother, and father all noticed the changes in him and questioned one another about Mitch's altered personality. He seemed to have changed abruptly, overnight, and Leah pinpointed the date to April 23rd. She remembered the night before, when he was tossing and turning, moaning, and at one point whimpering. She had intended to talk to him the next day about what had happened that night, but he left early for his office. Leah decided to talk with Lonnie about this massive change in her husband's behavior and focus. The two women met on a day when Mitch was out of his office, leading a campus demonstration.

Lonnie greeted Leah warmly and led her to a private conference room. "Lonnie, I want to talk to you about Mitch. We all know how much he's changed. We're all worried about him. He talks about how you've helped keep him stay sane and I want to thank you for that. Lonnie, he has changed. I don't understand what's gotten into him, but I do know that it all happened quite suddenly, literally overnight!"

Lonnie breathed a sigh of relief. "Oh, Leah, I hoped you would come in. You are absolutely correct. I remember it like it was yesterday. On April twenty-third he came into the office with a confused demeanor, but he also had fire in his eyes. He had a screaming argument with Dr. Cornelius and was asking about a graduate assistant named Casey, who never worked here."

"Casey, my gosh!" Leah exclaimed. "Mitch keeps yelling that name in his nightmares. And he keeps saying his name, too, when he's in his office at home. It actually seems like he's *talking* to someone named Casey. So, there is no employee, former employee, or friend that you know of named Casey?"

Lonnie shook her head. "I'm sorry, Leah, there is not. I even checked the database of science majors and there is no one with that name. Do you know anything about those series of dreams that Dr. Mitch had on the twenty-third of April?"

"I keep asking him and all he says is, 'There's little time left,' and 'Our future is in the balance.'"

Lonnie showed genuine concern. "May I ask you something personal, Leah?"

Leah took a moment. "Yes, Lonnie."

"Has Mitch sought therapy for his emotional distress?"

Leah looked down as a tear ran across her cheek. "Every time I suggest that, he turns away. It almost seems he isn't hearing me."

Lonnie looked thoughtful. She followed up with a suggestion. "Maybe we need to arrange a group meeting with Mitch; we'll call it a gathering of concerned family and friends. We're all very concerned about his state of mind, as much as you are. If he has an unthreatening gathering of friends and family showing concern, maybe he'll open up to us.

"It's a wonderful idea!" Leah exclaimed. "We'll do that and I would welcome your being with us."

Lonnie looked intently at Leah with a sense of wanting to express something else.

"Out with it, Lonnie. I know that look: You have something else on your mind."

"Leah, I have to ask—is Mitch taking any drugs? I mean hallucinogenic drugs. This whole dream scene sure seems like an LSD trip."

Leah smiled and touched Lonnie on her arm. "Not to worry on that score, my friend. Despite his unconventional teaching style and his aging hippy look, he's firmly against drugs. In fact, I overheard a father-daughter talk he had with Susannah last month—his passion on this is genuine."

Two weeks later, the gathering of friends and family took place at Mitch and Leah's home. Mitch was in his room and had not yet greeted the gathering of friends and family who had arrived a half-hour earlier. "I'm really worried," said Leah. "His door is locked and I've knocked several times but he doesn't respond."

Mitch was indeed locked in his room and shut out from the world. Around and around he went, his head spinning like a top, his thoughts swirling through his mind. *Am I here or back in a different dream? What the hell happened to Casey? I'm sure he was my graduate assistant. I still have the green glowing disc that he gave me when we parted. Did the dream change the reality of the present? Is this the present? What happened to Hurricane Cecil? I remember it distinctly as the last semblance of the present day before that branch hit me. Was the hurricane a dream or a dream within a dream? Oh, God where is Casey?*

Mitch heard thumping in the distance, interrupting his swirl of thoughts. It occurred to him that he been hearing the thumping for some time and now it was getting louder. Suddenly, he realized that he was in his home office and the thumping was on the door.

He heard the voice of his brother, Cal, shouting, "Last chance, Mitch. I will be breaking down your door in a minute!"

Is Cal real or is he taking me into another set of dreams? Mitch was suddenly shocked back into reality as something bounced off the door. Cal was trying to break it down. Mitch shouted: "Okay! Okay! I'm coming out." He opened the door and there was his big brother looking at him with tears in his eyes. Mitch moved to him and they hugged. He broke into sobs. "Oh, Cal. Cal. I need help. I can't take this any longer."

Cal took Mitch out to the living room where Leah, his father, Glenn, his friend Ray, and Lonnie were sitting. They all looked very shaken. Leah was crying.

There was an awkward silence. Mitch knew it was up to him to break the ice. "Thank you all for coming. I've been trying hard, but unsuccessfully, to sort through my mind for answers, but obviously I need your help and support. You are all the special people in my life. Dad, Cal, you've stood by me during this crazy time and other challenging periods in my life. Lonnie, you're so loyal to me. I wouldn't doubt you suggested having this conference. Ray, you've been my best buddy since grade school and you seem to be the only friend I have left. Leah . . . I love you and I want desperately to get back to being more normal again. I know I've been screwed up. I know I owe you all an explanation . . . but I just don't know where to begin." Mitch broke down, sobbing.

Leah came over and put her arm around him. "Honey, you need to start by telling us about that night of dreams."

"You'll all think I'm crazy." He smiled a little and said, "What the hell. What have I got to lose? I'm sure you all think I'm loony tunes already." Everyone, including Mitch, broke into spontaneous laughter and the tension broke from the room. After a few quiet moments, Mitch began describing his night of dreams. Periodically, he looked around the room. To his relief, everyone seemed absorbed in what he was saying; completely attentive to everything he was sharing. No one gave him a pitying look. No one was crying. It seemed that everyone had an open mind and was giving him a fair chance to describe what had happened to him. No one was writing him off as being psychotic or having had a major mental breakdown.

As he finished recounting his dream, he looked at every person in the room and openly expressed his fear. "I wouldn't blame you for thinking that this meltdown warrants my being put away. But I'm hoping you will all give me a chance to figure this out before you make a final decision. I do not believe I'm crazy. I believe something extraordinary has happened to me, so unique that it requires time to understand."

They were all momentarily silent.

Cal broke the pause. "Well I, for one, do not think that you are loony tunes." Everyone chuckled. "Mitch, you recounted your dream so lucidly and in no way am I certain that you are *not* a time flyer ready to take off again. It is certainly possible. I think that something dramatic occurred that night. Something that cannot be explained, and may never be explained. Tell us, what was the breaking point for you after the night of the twenty-second?"

Mitch thought for a while. "It was when Lonnie told me that Casey did not exist. He was so real to me in these dreams. With the risk of having you think that I am, indeed, psycho, I will tell you that he is *still* real to me." Mitch showed everyone the glowing green disc that Casey gave him during his farewell. No one laughed.

Now Glenn spoke. "Son, you have always been very imaginative. As a kid you daydreamed a lot. You have always followed the beat of your own drum. You always had crazy, very vivid dreams. I'm not discounting the possibility that what you described was something other than a dream. However, you need to consider the possibility that you have been under a great strain and your mind is giving you the message that you need to take a break."

Lonnie looked around the room. "Our department head, Alvin Klertz, is straight-laced and very authoritarian. Lately, I mean the period before the dream, he'd been riding Mitch to get off his unconventional teaching style of running his classes and start treating his course material in a more conventional manner."

Leah broke in. "Mitch, I've seen this strain building in you, but it's always been so hard for you to open up. You did tonight, and for that I am so grateful and relieved. I think for the first time since you were in the dream state, you are back with me again." She looked at Lonnie. "Tell us more about the work pressures."

Lonnie sipped her coffee. "Well, a rumor started that Klertz was trying to get rid of Mitch but the other professors in the department stood by him. The students loved his classes. Even the dean, who I know personally, was sympathetic to Mitch's situation, but he couldn't override the department head, at least not yet. In the meantime, Klertz turned up the pressure and demanded that Mitch become more aggressive in publishing and securing grants for pure biological research. Mitch was pressured but I sensed that he was holding a lot inside. Am I right, Mitch?"

Mitch sighed. "Yes, except for one thing you didn't pick up on. For some time now, I've been involved with the National Association of Academic Faculty Concerned about Climate Change. I've taken the issue very seriously and I strongly believe that we need to do something soon or we will lose our future, or I should say our children's future. Klertz knew that I'd been seeking National Science Foundation funding to create a national climate change education program that would demonstrate and involve students in a national climate-monitoring program with forums on the strength of the evidence and the message that we need to act now. Klertz thought that was too radical and wanted me to stop my production of these grant proposals and get back to the biological side of ecology. Of course, I refused and that ignited the strain between us."

Leah hugged Mitch. "Let's go away for the month. It will do both of us good in sorting this entire thing out."

Lonnie added, "And don't worry about Klertz. I'll talk to Dean Mason and get you a leave-of-absence approved right away."

"But, what about the rest of the semester?" Mitch asked.

"All taken care of. Dr. Fishbrook is ready to take over your classes on a minute's notice."

Mitch shook his head. "Lonnie, you really did think of everything. I'm going to surprise all of you and do this. I've always wanted to visit the Canyon Parks in the southwest. So, none of you want to commit me to the guys with the white suits from the loony bin?"

Everyone laughed again.

Glenn patted his son on the shoulder. "I, for one, do not think you are crazy. Let's vote!" Cal seconded the motion. Everyone said, "Aye."

But Lonnie added, "My 'aye' is qualified with the condition that you and Leah go away for the month."

Mitch looked around the room. "Yes, I will. Definitely."

Ray added, "Hey, I'd like to put a second motion on the table. I move that after all of this is behind us we encourage Mitch to write a novel about his dream experience."

Everyone laughed at this. Mitch felt a huge wave of relief and gratitude to the precious people in his life that stood by him. He was exhausted and Leah took him into their bedroom and set him comfortably into bed. In the recesses of his mind, he knew that the confusion was still there, but now he felt he could Let it Be. He went to bed with the Beatles' song "Let it Be" running through his head.

Mitch slipped into a deep sleep. He was visited by Casey, and his words were in his head. *"I'm sorry to intrude again and to have left you in such a state of confusion. We were going to wait a while to allow you time to process what you learned, but we need to get you back on the road to give you different perspectives so that you can understand more about the purpose of our travels. Then, you'll truly be able to Let it Be."*

Casey continued, *"I have new guides ready to take you on another adventure. You've met them before. Just ride with them tonight and you'll feel some peace of mind tomorrow. And go on that vacation with Leah."*

Mitch let go and drifted deeper into sleep. He knew that he was going on another journey, but he did not resist. Intuitively, he knew, as he fell deeper into the state of "lucid but weird" that he would be able to accept whatever was coming next with less confusion. *I must compartmentalize. Compartmentalize. And Let it Be.* Then he drifted even deeper into that state of weird lucidity with that same song still running through his head.

CHAPTER 4

THE GRAY DOOR (#4): THE SKEPTICS

What if the Skeptics are right?

***

2032

Mitch spiraled deeper and deeper into a spectacularly luminescent funnel filled with a bright rainbow of color. It seemed as though time had stopped and he was on an endless journey.

As trying as these last few weeks have been, I'm grateful I regained my balance; I do feel more prepared to meet whatever is facing me. The support my family and friends have given me over the past few weeks has been so helpful, and yet, now I've abandoned all of you again. But this is so beautiful! I feel my senses heightened—the glorious blues and reds mixing with bright yellows and greens are magnificent. I feel my emotions flowing from deep longing to exuberant feelings of joy and laughter. Am I dead? I feel so alive! This trip seems so different from the others. My mind is jumping from one thought to another. I'm awed by this new tunnel of time. Oh, Casey, what crazy place are you taking me to now?

Mitch suddenly broke out of his reverie, enveloped by a tunnel swirling to a bright spot in the distance. He passed through its small opening and once again found himself in the vestibule. He was startled, yet more alert than ever. *How am I ever going to explain this trip to anyone? Once was more than enough for them. This time, my family and friends are likely to think that I have truly gone off the deep end.*

These doors were different from those he had already passed through; they were bathed in soft luminescence he had never seen before. He was expecting Casey to come in at any moment, and shouted "Hey, Casey, these doors are the best ones yet!" and focused on the three doors facing him:

The Gray Door: (#4): The Skeptics
The Silver Door: (#5): Geo-Engineering
The Crimson Door: (#6): The Volcano

Mitch thought, *Okay, The Gray Door it is. Here we go again.* As he approached the door, he was suddenly greeted by a fellow who looked familiar, but whom he could not quite place. He was short, on the chubby side, and had bright red hair, possibly dyed. He wore a fluorescent body suit and talked a mile a minute. He greeted Mitch with a hearty laugh.

"Dr. Mitch, remember me? I'm Jake. I was there through The Green Door with my twin sister, Janey, when you arrived in 2044. Remember? We were throwing snowballs at you."

"Oh yes, how can I forget?" Mitch chuckled. "All I can say is that was a spectacular entrance I just came through." He definitely remembered them. In fact, he remembered every part of his journey through Door #3. Emboldened, he felt he had nothing to lose by being blunt. "I expected Casey, but I guess you'll be my merry guides here."

Laughing, Jake said, "Here, there, wherever and whenever we are. Oh, I forgot to tell you, it is 2032." Thinking Mitch was okay with all this, Jake turned around and walked toward the Gray Door as he rattled on.

But Mitch was not okay and he quickly interrupted Jake's monologue. "Whoa! Please! Stop right now! The last time we did a trip, it nearly knocked me out of my mind when I got back to my present. Unless you tell me what's going on here, I'm not playing, and I certainly won't go into any more futures with you!"

Jake stopped, turning in a dramatic pirouette, and gave Mitch a long look. "I'm sorry.

I understand." His head bowed slightly in a sign of genuine respect as he walked slowly back to Mitch. "Casey told me you would be a reluctant traveler this time, so let me tell you what I can at this point." He paused and took a deep breath while Mitch looked at him quizzically.

"Mitch, you are not in a dream. More important, you are not going insane or having a breakdown. Your present world of 2020 is intact and will be when you return. Unlike your last set of journeys, you'll return with complete peace of mind. What you will learn from our forays through the next three doors is how to compartmentalize your experiences so you can return to your world intact. I can tell you right now what type of reality you're experiencing, but you'll be able to accept it more readily and appreciate it more after we travel together through our futures tonight. Try to relax and trust me as we travel together through the next three doors."

Jake gave Mitch a few moments to regain his composure before he continued. Then, gently placing a hand on Mitch's shoulder, he spoke. "Most important, Dr. Mitch, the journeys we take will *not* continue endlessly. We will return you to your present-day life, fully enlightened with the lessons you will have learned. After tonight, you will be complete in your understanding of the different futures your world will face, depending on what it does to address climate change. When you enter the future scenes, just play your role. Don't worry about whether you'll be prepared to do so. You will be able to jump right in because you'll be part of that future. The only thing you need to think about at this point in your travels with me tonight is to have an open mind, a completely open mind. Give yourself peace of mind. Let it Be."

"But the time frame? In 2044, I'll be well over eighty years old. How does an old guy like me fit into this time-travel scenario? The next jump to a later future will have me over a hundred!"

Jake was prepared for this question. He pulled a mirror out of his sleeve and gave it to Mitch, encouraging him to look at himself. Mitch took the mirror and cautiously looked into it. What he saw was a forty-year-old version of himself.

"You see, Dr. Mitch, when you move forward in time, you not only absorb the knowledge of the future and all the years along the way, you retain the handsome good looks of your younger days. Now, can we get moving on?"

"Well, I hope I won't get hit in the head by a tree this time," Mitch joked in relief. Then, a serious look shadowed his face. "Okay, Jake. I will take you at your word. I'll trust you and this journey to whatever dimension you take me to, and I'll try to free my imagination and learn from whatever I see. I'm at your mercy, Jake, so don't screw it up. Let's go."

Mitch and Jake passed through Door #4: The Skeptics. They were swept up into a swirling passage that dropped them with a *sudden pop* into a new reality—an entire world that seemed stained with a subtle shade of gray. Even the blue lake was tinted gray, although a few features of the landscape—like the leafless trees in front of him—were a more vibrant gray. It was as if he had entered into a black-and-white photograph. *I'm not going to think about this too much. I accept the grayness of whatever reality this is*, Mitch told himself.

"We need to get going. I need you to see this documentary." A patch of wavy air appeared immediately above their heads. Jake opened his transponder and waved it up, across, and down. A screen appeared in the air in front of them, and out of an increasing resolution of pixels, a newscaster appeared. He was reporting on the final results of the research initiative of the last ten years.

"Good morning. I am Howard Hanaford, reporting from our National Univision News of 2032. We're here today to discuss the results of the findings issued by former President Mazikowski's 'Climate Change or Not' (CC/N) initiative."

Hanaford paused and took a sip of water. "First, some background. The most dramatic circumstance since 2020 was the unanticipated reversal in climatic trends. In the early 2000s, despite temperatures continuing to rise, the rate of increase began to slow, as did the rate of global warming.

Then, starting in 2020, temperatures began to stabilize; by 2024, they began to decline. These seemingly contradictory events happened despite the fact that CO_2 emissions continued unabated. This precipitated a change in thinking by climate science scholars who published in major peer-reviewed science journals.

"Thirty years ago, 97 percent of scientists' published studies supported the idea that the climate was warming and human emissions were the cause. Now, in 2032, the number of peer-reviewed papers supporting climate change has declined to 48 percent—less than half! However, many of those studies cite the fact that the recent temperature decline is only a temporary slowdown, and predict that the world is bound to start warming again. On the other hand, the 52 percent of scientists who have published studies now question the entire theory of climate change, asserting that we still do not understand enough about how the earth's thermostat works to justify mandated actions to reduce carbon emissions. From near certainty to a fifty-fifty split in scientific opinion in a mere eighteen years is astounding! And, as far as the general public is concerned, only 20 percent believe that the world will start warming again. As we know, in the end public opinion outweighs the scientific method in driving policy."

Hanaford paused to allow this dramatic change in the climate trend over the past decade to sink in. "With open-mindedness, scientists recognized and verified these changes and rejected their near-unanimous conclusion in 2014 that 'climate change is unequivocal.' Some of the former skeptics –now in the mainstream-- even compared these radical and unpredicted changes in climate warming rates to the certainty of the scientists of the seventeenth century who believed the sun and other planets rotated around the earth, only to be turned around by the science of two skeptics: Copernicus and Galileo. Nevertheless, the ultimate decision of what to do was politically driven, influencing the decisions the president and Congress would be willing to fund: immediate mitigation legislation, or a new study to determine whether the old climate change model was incorrect, or both.

"In 2022, President Mazikowski ordered a comprehensive ten-year study to answer, once and for all, whether climate change was occurring, whether human emissions of CO_2 were the cause, and what future outcomes will be. President Mazikowski faced a barrage of criticism for ordering this initiative by advocates who promoted instituting 80 percent cuts in greenhouse gases immediately. This included environmental activist groups, many mainstream climate scientists, and a handful of industry leaders. The Blue Sky Party organized a massive campaign to stop the research study. Its members rejected what they thought of as an unnecessary repeat of the IPCC studies of 1992–2015, which had concluded with a loud 'it's happening' message in 2014.[20]"

Hanaford paused to take another sip of water, and then continued. "But in the end, Mazikowski took advantage of the Citizens' Unity Party's domination of Congress to get his proposal through. Now, we are at the end of his proposed ten-year research program and their final report has been issued. Let's join the discussion of six scientists from major research centers who had input into the research project." Suddenly, Mitch found himself pulled into the screen: he was participating in the discussion.

Hanaford opened the discussion. "Dr. Shelski, what do you think? The report did not seem to give a yes or no as to whether climate change is happening."

Shelski paused a moment before replying. "First of all, as an environmental policy analyst, I can never expect science to give a clear black-or-white answer. We interpret our findings in probabilities of significance. As you said, the views of mainstream scientists have flipped from near certainty to fifty-fifty in a mere decade. We need to know more. The president's proposed ten-year study about whether and how to mitigate was the way to go."

Hanaford moved the discussion along, asking: "What about temperature rise. It is still happening, isn't it? As a climatologist, Dr. Bertrand, what do you think?"

"No. There has been a slight decline over the past few years."

Dr. Shelski broke back into the conversation. "You need to take the oceans into account. What we did not expect, twenty years ago, was that the oceans' rate of absorption of CO_2 has increased. This remains unexplained, since the common scientific perspective holds that when oceans warm up they normally absorb less CO_2.[21] But we now see that the opposite is happening: the oceans are continuing to warm, but our global temperatures are decreasing. That means a good portion of the human-caused emission of CO_2 is being soaked up by the oceans' sponge more efficiently than ever before."

Mitch broke into the conversation. "Dr. Shelski, have you seen the latest studies in *The World Climate Change Journal*? One cites the fact that the absorption of CO_2 by the oceans has reversed itself since 2031. It is now starting to release CO_2 into the atmosphere. A second study, published in May, corroborated this result and proposed a theory that explains the twenty-year increase in CO_2 absorption from the oceans and predicts a new phase of CO_2 release. We will be getting a "burp" from the ocean, which will make up for these past few years when we were being protected by the oceans' sponge effect." Mitch paused to allow for a response.

---

20 See https://www.ipcc.ch/pdf/assessment-report/ar5/syr/AR5_SYR_FINAL_SPM.pdf
21 See http://news.wisc.edu/climate-change-reducing-oceans-carbon-dioxide-uptake/

"Dr. Shelski, do these studies contradict the findings of the ten-year research initiative?" Hanaford asked. "No, not at all. These studies came after the CC/N report was issued. As I said before, science does not give black-and-white answers. I read both papers in The World Climate Change Journal and they seem legitimate in their technical findings. We need to watch that trend and put more funding into ocean absorption research. However, this does not abrogate the findings of a decade of CC/N research. We are still seeing that the "consensus wave" of the scientific studies has flipped toward those who are skeptical."

Hanaford posited to the entire panel, "Are we saying that those from a decade ago who claimed that climate change is a hoax may, in fact, have been right?" Dr. Carl Bertrand interrupted. "I, for one, do not see it that way. I have always drawn a distinction between the 'contrarians' and 'the skeptics.' The contrarians were saying, 'No. Not possible. The climate is not changing from human-caused emissions. Period.' The skeptics, on the other hand, are legitimate credentialed climate scientists who questioned the details and the weight of the evidence, but kept their minds open to the possibility they may not have been correct. The skeptics of fifteen years ago have morphed into today's mainstream, while those of us who still believe that climate change is happening—although perhaps not in the manner we once thought—are increasingly being viewed as the skeptics. The debate continues."

Turning to Mitch, Hanaford asked, "Do we know for sure that the evidence from the CC/N report is definitive?"

"We are not absolutely sure. As Tom said, we will never have a 100 percent certain proof of anything we're studying. In 2014, 97 percent of us believed that the world had become warmer due to human-caused emissions that were causing serious negative effects on society. After ten years of CC/N research, it appears the views of half of those scientists have changed. But I have serious questions about many of those CC/N studies. I believe those who were skeptics in 2014 voiced doubts on the global warming perspective and the research studies pursued the path of proving those doubts as correct. Despite their decade-long study, however, I contest its conclusions. The report simply strove to verify its own assumptions. For example, the recent findings of CO_2 release from the ocean, challenge—indeed refute—one of the cornerstones of the ten-year CC/N research. I believe these ocean release findings will begin to unravel the findings of the entire CC/N report."

Calvin Curtiss became agitated "Hogwash, Mitch! You environmental alarmists have been crying wolf for fifty years. Now, when the consensus of mainstream scientists has shifted in response to new evidence, it's you people who are being the contrarians. As a civil engineer, I stand for finding solutions to problems through sound science. You guys are not speaking with one voice."

Mitch's cheek twitched and he banged his fist on the table, getting everyone's attention.

Wait! He told himself. *Slow down, keep an open mind. You'll have your chance to offer your response. For now, open your mind, speak calmly and listen to others. Let it Be.* He regained his composure with a calm he didn't know he could summon. "Okay, Curtiss, I grant you that there are new findings and I need to take that into account as I explore my objections. But please, do not label me a contrarian. Think back on Dr. Shelski's distinction between contrarians and skeptics. In terms of the present-day debate, think of me as a *new* skeptic. Now, let's agree to disagree and move on."

"Thank you, Mitch. Well said." Hanaford added, "This discussion needs open minds and civil discourse. Now, let's turn to the solutions. Is mitigation of our CO_2 emissions possible? Is it necessary?"

Curtiss smirked. "Well, you all know I stand by the CC/N report. I do not think we need to mitigate at this point. However, in the spirit of open-mindedness, I'll allow that perhaps we should continue researching and monitoring climate and, in the meantime, invest in mitigation options in case the findings in the next ten years shift the pendulum of consensus again. I yield the floor to Dr. Marcus on this subject."

David Marcus entered the discussion. "Thank you, Calvin. As you all know, I'm an environmental engineer and have been involved in the national research program to explore geo-engineering fixes. I know that some of you may consider this reckless, but I look at having a quick solution in hand as cheap insurance for the possibility that the 'new skeptics,' those climate scientists who are still convinced that the world will resume overheating in due time, may have been right all along." He looked at Mitch. "We would not intend to deploy it, but if the new skeptics are right, we would. Deploying a geo-engineering fix would be reserved for the moment when we had no time left to solve a climate emergency through reducing emissions. It could be a temporary stop-gap while we reorient society to a carbon-free culture, which I still believe is a necessary thing for society to do in any event. As you know, President Rother is now proposing a second phase of research she has dubbed the 'Mitigate or Not' (M/N) initiative, aimed at examining all of the alternatives for taming global warming. At the same time, the M/N phase of research will continue to study new analyses of climate data to help us better define and understand the earth's natural processes of modulating the dynamics of the climate system as our planet continues to accumulate $CO_{2 \text{ in its' atmosphere}}$. But getting back to your original question, Dr. Hanaford—there are solutions already out there if we need to mitigate. As you know, I have devoted my career to developing the geo-engineering solution of deploying a fleet of unmanned spacecraft to orbit our earth and inject reflecting nanoparticles that would deflect incoming ultraviolet and visible radiation, thereby reducing the amount of heat radiation to our planet. Of course, the technology I've developed and all other possible mitigation solutions will be subject to intense environmental and economic scrutiny in the M/N research over the ten years, if it is funded."

As Dr. Marcus spoke, Mitch felt a slow burn of anger that accelerated to a tempest of rage. His cheek twitched again and his face began to redden. He wanted to pound the table and shout out his objection. This time, the twitch caused him to pause. *I need to calm myself and open my mind. But I need to say something here!*

"Respectfully, Dr. Marcus, I object strongly to the whole idea of geo-engineering.

Reducing the amount of solar radiation will tamper with our relationship to the sun—and the most fundamental control of our climate system. It would be like unleashing a herd of wild horses from a corral. Once done, we may never get control back. If you reduce radiation for a period of time, the consequence may be to somersault our climate into a new, self-reinforcing, state of disequilibrium, such that if we wanted to take the spacecraft out of orbit, the damage would have already been done. Having said that, if the nation sets a course of shifting the emphasis toward mitigation, I would want to see the environmental/economic impact analyses of the geo-engineering solutions you've suggested as well as an economic and environmental impact analysis of phasing in 80 percent emission cuts."

The session wound down and everyone shook hands. Mitch felt himself falling out of the screen and rejoining Jake in the gray-tinted meadow. He looked at Jake and thought about what had just happened. Not the debate itself, but falling out of what we used to call a monitor or television screen and landing back where he had come from. "Jake, how did I do that? One minute I'm watching this crazy screen appear in thin air, then I'm transported into the discussion, and now I'm back."

Jake smiled. "Technology, Dr. Mitch. Let it Be at that. Remember, this is 2032 of another dimension. Suspend your present and let your 2020 imagination fly free. The important thing is that the transponder-based communications era allows anyone, anywhere, to immediately tune in to important issues of our times. In fact, you do not have to 'turn on the television' or 'log on,' to computers as you did in your 2020 dimension. When the broadcast airs, waviness appears in thin air around every person on the planet. You can turn it off simply by walking away. If you want to watch, you wave your transponder back and forth, up and down and the wavy air is replaced by three-dimensional images, like the debate you just participated in. That is your active participant role in these journeys. By the way, all broadcasts are educational without the commercials you had to endure back in 2020. We even vote that way—87 percent of our population now participates in electing our leaders." Jake ruffled his bright red hair and shook his head in his usual peculiar way.

Mitch focused on Jake's eccentric hair and became irritated at this odd fellow. He missed Casey more than ever. But he shook his head and dismissed the quirks of this time guide, moving on. "Let's talk about the debate. That was surreal. I know you warned me I would be able to participate with the knowledge of the events of this future, but it just felt weird."

Jake replied with a little irritation of his own. "Let's focus on what's really important, Dr. Mitch. What did you think of the discussion?"

Mitch took a deep breath. "I've been processing it while we were talking about these other things. I can't say I am happy with where this age is headed. I suppose that our science of the earlier part of the century had convinced me that climate change was happening and imminent catastrophic outcome was inevitable. I still believe we are changing our climate for the worse. But perhaps we need to hedge our bets about the outcomes we were predicting. As I said in the debate, even though our temperatures are now declining, I still firmly believe the warming will return. I still think we need to wake up and accept that inevitability."

"What were you feeling during the debate, Dr. Mitch?"

Mitch sighed. "Weird. I felt my anger bubbling up at Curtiss and Marcus and then felt something pulling me back from fanning the flames of my rage. That's not like me, but I must say it was refreshing. I suppose you're going to tell me more about that harnessing of my anger, Jake."

Jake smiled. "You did well, Dr. Mitch. You're learning how to live in these dimensions. You are letting it be. Now, are you ready to go on?"

Mitch felt a calm wash over him. "Yes, where are we going now?"

"We're traveling to 2042, in the same channel of Door #4, but in a different time period. Are you ready?" An orange screen appeared in front of them and Jake jumped in with Mitch following behind. As the green-gray meadow began disappearing, Mitch saw the Beatles standing in the long waving grass, singing "Let it Be."

2042

Mitch and Jake found themselves sliding down an old-fashioned playground slide. They landed in a desert, on a ground of coarse dirty sand. Mitch sensed that they were not yet in Door # 4, but were at a stop-off to pick up another passenger. And there she was. . . Janey, in a tight-fitting spandex suit that matched Jake's bizarre outfit, ran over to greet them. She looked much better in her spandex and still had the same shade of bright red hair. She observed Dr. Mitch admiring how very sexy she looked and she smiled at him with a gentle wink.

I don't get it. How could Janey be attracted to a long-haired leaping gnome like me? Then he remembered his reflection in the portable mirror that Jake gave to him. *Oh, I see. Being transformed into a young dapper stud is a nice perk on these journeys.* Mitch shook his head and wondered whether this was simply a dream. He shook his head one more time and told himself: *Suspend your imagination, man. Enjoy this ride.* With Janey in the picture now, it seemed to him that this journey wasn't so bad, maybe even fun. So he went along and decided to keep his mind wide open.

Janey and Jake hugged. "It's about time you arrived here, sister." Mitch remembered that Jake and Janey were twins.

She smiled. "I decided to let you take Mitch along on your own for the first leg. How're you doing, Dr. Mitch?"

Mitch gave her a thumbs-up. "I'm learning, and trying to go with the flow."

"That's great," she replied. "That means you're getting the hang of it."

"Okay, where to now?"

Jake took out his transponder, waved it up, down, and across, two times. The world turned from sand-brown to arctic white. They were in a whiteout, fully dressed in arctic snow gear. Their bodies were enclosed in layers of surprisingly thin, warm, shimmering synthetic material, and insulated boots and gloves that seemed natural extensions of the colorful suits. Clear synthetic, semi-soft bubble helmets encased their heads. Amazed, Mitch noted that his bubble didn't fog up as he breathed.

Walking was hard. No matter how hard he pushed forward, Mitch felt the power of the wind blowing him sideways, and in only minutes he felt exhausted from the effort. Suddenly, as he struggled to keep up, his helmet was filled with Janey's voice.

"Are you okay, Dr. Mitch?"

"Roger that. It never ceases to amaze me how much energy your body puts out in arctic conditions."

Jake's voice came back with his characteristic laugh. "You mean Antarctic conditions! That's where we are."

Then Mitch's helmet filled with a new voice.

"Okay, you three, enough socializing! Welcome aboard, Dr. Mitch, this is Captain Holbrook. Get yourselves to our command vehicle so we can prepare for our trip to the Ross Shelf and retrieve the ice cores we've come to get. Come on in out of the cold."

A red-and-white-striped vehicle appeared on the horizon. The craft opened its sliding doors when it arrived. It appeared to Mitch to be a type of hovercraft, but there was no wind-thrust noise, just a slight whirring sound

As a gull-wing door opened, Mitch, Janey, and Jake entered the vehicle and shed their winter gear. It was pleasantly warm and comfortable inside. The hovercraft was huge, the length of the old 747's fuselage. There were banks of refrigerators lining the walls and laboratory benches filling the center. A stove sat in the middle of the floor with hot chocolate, coffee, and tea waiting for them alongside three translucent boxes. "Help yourselves, folks," a mechanical voice emanated from the boxes. "I am Bart, an assistant to Captain Holbrook. He will be joining us shortly."

Mitch stared at the box. "What gives with this?" The three others laughed.

"Wave your hand in front of it, Dr. Mitch," Jake instructed. When he did, a menu of food selections appeared.

"Okay," chuckled Mitch, "I'll go for it. How do I get the slice of pizza with everything on it?"

Janey reached over, waved her transponder over the box, and a keyboard slid out. "Just tap the letters on the keyboard, like your old computer devices from the earlier part of the century." Mitch tapped the keyboard and almost instantaneously, two slices of hot pizza popped out of the translucent box. "Delighted!" he said, devouring the food as Captain Holbrook entered the main area from a door near the front of the hovercraft.

"Okay, team. We're almost ready to take off for the Ross Shelf."

"Captain, I'm Mitch Warner from the Miami Environmental Research Center. I have a concern about going out there if it's about to calve off."

Holbrook stared at Mitch. "Yes, Dr. Warner, welcome! I know who you are: the scientist from the other dimension. Please call me Holbrook. We have covered the shelf with spectrographic monitors, which measure the adhesion of the ice shelf to the continent. We have weeks, maybe months before it happens, so rest assured we will be safe. Now, let me get this mission started. We have about forty-five minutes travel time, so we can chat while we glide there." Holbrook left the room, moving toward the front of the craft, where Mitch presumed the operating controls were located.

Mitch approached Jake and Janey with a look of confusion. "I am trying to be comfortable with all of this, but being in Antarctica is disorienting. For one, why is the Ross Shelf ready to peel off with the pause in warming findings of the CC/N report?"

"That was ten years ago, Dr. Mitch," responded Janey. "A lot has happened since that time. The president's M/N package was funded by Congress back in 2032 and you'll be happy to know that Mazikowski's successor, President Rother, included a chunk of funding for enhanced monitoring of climate change. Here comes Captain Holbrook—he can explain what the world learned over the past decade. Mitch looked at Holbrook approaching and wondered: *Who is piloting this craft*?

Captain Holbrook re-entered the main cabin, sat down next to Mitch, and asked if he had enjoyed his food. "Not bad for a meal that popped out of a translucent box," Mitch said, graciously.

Holbrook turned serious. "We are on a mission to retrieve four ice cores from the Ross Shelf and bring them back to the central station for analysis. As a climate ecologist, you need to ensure there is nothing in the extraction process of the ice cores that could compromise their integrity for analysis. These ice cores will be long[22]—over three miles deep. They should give us the most comprehensive view of past climate history achieved to date, as we expect to date the bottom of the core to approximately one and a half million years. We only recently developed the drilling technology that enables us to go down that deep and there is a lot of excitement in the scientific community about what we might learn about past climate fluctuations. Your responsibility is to monitor how the cores are handled: everything from the procedures to prevent break-up of the ice cores to the handling and transfer of the cores back into the hovercraft—and then the transfer to the lab at the Vostock Station. You will document each step of the extraction, transport, and lab transfer procedures. Perhaps you could brief Jake and Janey about what ice cores are and how they are used to infer past climate history. So that, Dr. Mitch, is your briefing. Any questions?"

"I'll be happy to brief Jake and Janey, no problem. And yes, I do have questions—actually two," Mitch replied "First, why did you pick the Ross Ice Shelf for the coring?"

"The ice is deep enough here," Holbrook responded. "More importantly, the analysis of ancient climate conditions proximate to the ocean may give us a better picture of past climate fluctuations. As you know, the interaction between oceans and land masses forms a significant part of the ancient world climate picture. In the sections of the core that we analyze, we'll employ the 'ice/rock adhesion' test to predict more precisely the calving of the shelf into the ocean. What's your second question, Dr. Mitch?"

"Who is driving this thing while you're here talking to us?" Captain Holbrook smiled. Then Mitch asked his second serious question. "As you may know, I just skipped in from 2032." Holbrook looked confused at this, but let it pass.

Mitch wondered about this. *Why would Holbrook be confused? He's already said he knew who I am and where I came from.*

Holbrook continued, "Back then, Mazikowski's CC/N (Climate Change or Not) report had just been released, demonstrating that the ten-year study of climate change had concluded there was some doubt about whether the climate is changing and leading the world to a devastating future. The consensus of mainstream scientists in 2032 was that there is more doubt than confidence that human-induced global warming is significant. This was a virtual flip from 2014 when most mainstream climate scientists believed that the evidence proved that climate change was real."

Mitch questioned this information. "I'm still in the latter camp, so I guess I'm the new skeptic to the mainstream of scientific thought. In 2032, we were ready to launch Phase Two—President Rother's initiative to study mitigation procedures while continuing the

---

[22] Currently, ice cores extracted from Antactic Ice are almost 2 miles deep (3,000 meters). See http://www.antarcticglaciers.org/glaciers-and-climate/ice-cores/ice-core-basics/

study of climate trends. My question is what were the conclusions of the Phase Two study? It seems to me that if the Ross Ice Shelf is ready to calve off into the Southern Ocean, we must have returned to a warming period."

Captain Holbrook smiled appreciatively. "How much time do you have? That's a lot of material to cover in a half an hour, but I'll do my best to summarize. "First let me clarify: the entire Ross Sheet is not in danger of calving into the ocean; only a big chunk of it. The main thrust of the CC/N study was to examine costs, benefits, and impacts of alternative mitigation fixes. This turned out to be focused mainly on geo-engineering. I'll defer that question to your own review of the report after this mission is complete."

Jake whispered to Mitch, "Maybe there's another journey in your future to a dimension that deployed one of the geo-engineering fixes."

Holbrook continued, "As for the 'warming or not' part of the research, it isn't so cut and dry. Yes, the Antarctic is getting warmer, as is the Arctic. The evidence of this part of the Ross Ice Sheet coming apart came to our attention about five years ago and the U.N. channeled international funds to extract as much information from the ice as we could before it sheds into the sea. However, other than the poles, the planet is still in a period of a gradual temperature decline. This trend has been observed since 2020. Most people—65 percent in the latest global poll—believe the recent decline in temperature is here to stay and the global warming theory is out the window. But clearly, the last twenty years have shown us there is something different happening in our climate system that is not in line with the climate change theories of the first couple of decades of the twenty-first century. All the models from that period have been proved wrong. Clearly, we did not—and still do not—have as clear an understanding of how our climate system works in assimilating increases in CO_2 from world emissions as we'd presumed. On the other hand, average annual CO_2 measurements continue to show the same trends as the Keeling Curve of the 1950s: a gradual but steady rise in CO_2 in the atmosphere year in and year out. Also, the concern that you raised in the 2032 Marcus Holby broadcast regarding the oceans'

release of heat has been verified, corroborated, and accepted as part of the climate picture. Since 2030, the oceans have been pumping heat and CO_2 into the atmosphere instead of absorbing it out of the air. This is confirmed by the recent increase in the rate of rise of CO_2 release from the oceans into the atmosphere since 2032. So, the CO_2 levels have increased to nearly 460 parts per million (ppm), passing the 450 ppm benchmark that climate scientists from the early 2000s noted as a critical point-of-no-return tipping point. Back then, they thought that amount of CO_2 would trigger feedbacks that would accelerate temperature increase and lead us, ultimately, to devastation. So, we have a conundrum here: CO_2 levels have increased past the critical 450 ppm threshold, but those adverse feedbacks we expected haven't yet occurred and temperatures are still decreasing. So, here in Antarctica we are striving to add one more piece of understanding to our knowledge of how our climate system really works. It seems to me that we almost had the understanding of climate mechanics nailed twenty-five years ago, but I guess we were missing something. We're still scratching the surface of the truth of the matter."

Only the slight whirring of the hovercraft drive interrupted the silence that filled the room. "Oh," added Holbrook after a moment. "To answer your other question, don't worry about the hovercraft. It drives itself with a triple-modulation sequencer array." The captain smiled again.

Mitch took a deep breath. "You see now, Captain, this is where my angst and confusion kick in. Twenty-five years ago I was on the forefront of those who claimed victory in our understanding of human-caused climate change. We in the scientific community had achieved a consensus on the matter and were urging society to move on and mitigate. The few skeptics out there kept nipping at the edges of our virtually proven theory and slowing us down, claiming we needed better evidence and a rock-solid understanding of climate before we mitigated. Now, in this world of yours, *I* am the skeptic. I have a lot of catching up to do. I should read the entire publication of the M/N report carefully, with an open mind, to see if it changes my perspective on whether we need to mitigate now (M) or Not (N)."

"Actually, Mitch, although the public is still lulled into the 'global warming is passé' mindset, the consensus of mainstream climate scientists is beginning to swing back to your view of reality."

"One more question. Why are we going out in the blizzard to retrieve the cores? Why don't we postpone until weather is safer?"

"The coring team has been out there for a long time and they have informed us that the blizzard conditions may push some of them into dangerous situations. They already had one member suffer hypothermia. So now it is more of a rescue mission."

As much as he wanted to continue the conversation, Mitch felt overwhelmed by all this information and thought: *This is enough for now.*

Captain Holbrook seemed to read his mind. "Mitch, that's a reading task for another day.

Now we have to focus on the project at hand: we're nearly at the Shelf. Get your gear bags ready, everybody. We're only minutes away from facing the blizzard outside." He turned to Mitch. "Remember! We need to retrieve four cores and you're our quality-control officer."

"Aye-aye, sir!" Mitch saluted Captain Holbrook as the captain returned to the control center.

As the craft began to slow to an approach speed, everyone noticed that what had been a light snow was now thick and falling heavily, turning into a storm. Mitch, Holbrook, Jake and Janey were grateful for the protection their special blizzard gear would assure, as they prepared to suit up.

They were gathering near the exit door when Captain Holbrook interrupted, telling them one more team member would be joining the mission to the drill site. A tall man with curly red hair and a matching beard came out of the front room. He looked vaguely familiar to Mitch.

Jake and Janey yelled out in unison, "Charley!" as they ran to him. Charley hugged Janey and he and Jake slapped hands and bumped fists. They broke out in identical laughs.

Oh no! I guess that I will have to just get used to two of these jokers. Mitch looked at the three of them and realized why Charley looked familiar. They were related. "Let me guess, Charley is your brother," he said to the twins.

"Close," said Charley. "I am their cousin and soul mate. Do I have time, Captain?" he asked.

Holbrook seemed to know what was coming. "We have ten minutes. We could use some levity before going out there."

Charley pulled out a banjo from his gear bag and entertained the group for ten minutes, alternating between singing songs and telling short, amusing jokes. *He is most certainly an entertainer and seems to be a real nice guy*, Mitch thought, grateful for the interlude.

Holbrook broke in. "Sorry to interrupt the festivities, Charley, but now it's time to suit up. When you're all ready, gather by the door."

"Okay," said Charley. "We'll continue with festive celebrations tonight!"

Minutes later, Captain Holbrook was outside, guiding the team down the ramp of the hovercraft. Mitch stepped into the deep snow, wearing his newly issued hover-powered snowshoes. He was happily gliding on the snow with each step instead of trudging through it. Jake and Janey were behind him. The blizzard raged in their faces. Visibility was near zero. "Dr. Mitch," Janey offered helpfully, "push the green button on your left shoulder to activate your climate control."

Jake's voice filled his helmet. "What do you think, Dr. Mitch?"

Mitch turned around and tried to see the twins. "I think I can barely see."

Janey came on the radio. "Dr. Mitch, press the large button on your right shoulder. That will give you an infrared assist for your vision."

Charley came on and joked: "Oh, and Mitch, when you press the button, your irises will turn red and your eyes become the flashlight. Don't worry, though, as soon as you turn the button off, your eyes will return to their normal color and function."

Captain Holbrook broke in. "All right, team, listen up. Enough horseplay. Charley can return to his antics when we get back to the craft after the mission. Now we need to be serious. We need to lash ourselves together." He handed the carabineer of the tether to Mitch, who passed it to Janey. "Keep communicating and every five minutes do a buddy check. Ask the team member behind you to verify he or she is still there. Don't bother looking behind you; you will not be able to see in this blizzard."

"Why didn't we just drive the hovercraft right up to the drill rig?" Mitch asked.

Holbrook shook his head. "Too chancy. There are chasms here that will be covered over with snow."

"How are we supposed to know where these snow-covered chasms are? I don't fancy falling into my own grave!" Mitch shouted.

"Easy, Mitch," replied Holbrook. "We have our route mapped to avoid those chasms. We are all equipped with GPS trans-depth meters that will sound an alarm when we approach a chasm or when we step off track. Now, press the large button on your helmet just above and between your eyes." Mitch did just that. The visor of his helmet turned into a map marking their route with a green X from where they were currently standing. Zooming out, it showed the location of the nearby chasms. "That is fantastic!" Mitch shouted back. They all laughed.

Holbrook continued, "If you detach from your tether and get lost, that should help you get back to us. Otherwise, just follow the tether. I will be in the lead and will be monitoring the same map as we proceed. It's only about three hundred meters to the drill rig and we should be there in half an hour or so."

And all I have to do is keep tethered and follow the GPS route, Mitch repeated to himself, realizing that a hundred-yard walk trudging through deep snow on a clear day may take only ten minutes, but even sliding through the blinding blizzard on their hovering snow shoes, fighting the wind, would slow them down to a crawl.

There was one other question he needed cleared up. "Holbrook, how did the drill rig get out here if it's too risky for our hovercraft?"

"Our hovercraft, designed for light transport. It is suspended a foot above the ground. At that distance, our craft can break open a chasm through its undercarriage heat, with the risk of pulling us all into the abyss. The drill rig we are headed toward is specially designed to be a flying and landing vehicle, something like a large helicopter. It does not move laterally across the ice and snow. Unfortunately, they lost their engines on their descent; so in addition to the hypothermia conditions, this is also a part of the rescue mission."

Now Holbrook became impatient. "Now, if there are no more questions, let's begin and follow protocol. There are only two communications I want to hear, either the head-count or a warning that you are in trouble!" Everyone was silent as the captain continued. "Our first count starts now . . . and so far we have been sloppy in our communications. Conclude each with 'over.' Okay. Let's go. Over." Captain Holbrook started the first count. "Mitch. Over."

"I'm here. Over."

"Janey? Over."

She replied, then Jake, and finally Charley. The first check was complete. It took almost forty-five minutes until they saw the lights of the extraction crew at the drill hole. Mitch noticed a number of special sleds on site, apparently the vehicles that would transport the cores to the ship.

"Holbrook, here. You should all see the site now. Let me summarize the protocol that we reviewed in detail on the hovercraft. As soon as we arrive, the extraction crew will start transferring the cores to the sleds. Mitch, you get right in there to observe. Jake and Janey, assist in the loading of the extracted cores onto the sleds. Mitch, you need to sign off on your manifest at the extraction of the cores and their transfer to the sleds. When all the cores have been loaded, we wait for the crew to pack up their gear, and walk back to the hovercraft with them. Then you need to observe and sign off as the cores are carried up the ramp and into the hovercraft. Jake and Janey, you need to do all of the handling of the cores and make sure Mitch is not distracted. He must remain focused on recording accurately all phases of the entire procedure. I know I've gone over this many times, so we should all have muscle memory of our roles. It has taken five grueling months for these guys to get the cores out of the ground and we do not want to see any of them disqualified due to contamination. Ready, team? Over."

Mitch followed up. "Aye-aye, Captain, but one question. How do I verify that the cores were taken out of the ground without contamination? Over." Mitch was knowledgeable on coring technology, but unaware of the most recent advances in extraction operations on the ground.

Holbrook responded, "The extraction process is fully automated and pulls the ice out of the holes in a double-encapsulated Teflon tube. The annular space between the tubes is suspended in a vacuum. This will be the first time the cores are handled by human hands. The cores are transferred to sleds that glide above the ground like mini-hovercraft. That is where you come in. Over."

Mitch felt excited to be a part of this team effort and delighted by the incredible technology. *How am I ever going to be able to go back to the stone age of 2020?* Minutes later, the team met with the extraction crew. A radio transmission came through their helmets. "It's about time you got here, Holbrook. Over."

"Holbrook here. Greetings, Constantine. Team, raise your hands as I introduce you. Behind me is Dr. Mitch Warner, red suit. Behind him is Janey Smith, gold suit; behind her is Jake Smith, blue suit. Bringing up the rear is Charley Addison, green suit. We'll save the introductions of Constantine's crew members until after we make it back to our craft. For now, all of your communications are between each of us and Constantine. Over."

Constantine came back on. "Smiths, are you related? Over."

Both Jake and Janey came on at the same time saying, "Twins." Jake broke into his distinctive laugh and they both said "Over."

"ALL OF YOU!!!" Holbrook barked. "No more small talk. Save it for Charley's party tonight."

As they approached the drill holes, Mitch looked around. "Where's the rig? Over."

Janey responded, "It's all here. Portable on a hover float deck to allow a remote drilling and pull out. In the old days, it took around one year of drilling to extract a 3,000 meter[23] core. With this new technology, we pulled this 6,000 meter core out in a half year. After they're done, the entire apparatus will fold up into a compact package. We should be out of here in an hour. Over."

Holbrook cut in. "NO TIME DEADLINES HERE! We will stay out here all day and all night if necessary, to get this right. They have portable shelters on site for us to escape the storm when we need to. And, of course, the drilling team has temporarily taken haven here and they still have a two day supply of food. No one should feel any pressures. If any of you, particularly you, Mitch—needs to slow up, yell, 'stop.' Understood? Over."

Jake and Janey said in unison, "Aye-aye, Captain. Over."

Mitch followed up, "Got it. Over."

Charley had the last word. "Okay, here. When you get antsy for quicker action think of my banjo playing 'Slow Down, Pokey Boy.' Over." For whatever reason, Holbrook gave Charlie a pass on the small talk.

It took one hour and twenty minutes to complete the operation. Mitch signed off on six different steps of the process, verifying there was no human contact or breakage of any of the cores. Captain Holbrook led the return to the hovercraft. Jake was behind him, in front of the five sleds being pulled by the extraction crew. Mitch moved among the sleds to ensure there was no contact or breakup of the cores. Janey assisted Mitch. The sixth member of Constantine's crew walked behind the sleds, monitoring the transport of each core, while Charley stayed in the rear of the caravan, watching out for the entire team. The return to the hovercraft went smoothly, until the last buddy check. Charley did not respond. The blizzard was blowing harder than ever.

Holbrook broadcast: "Who was he tethered to? Over."

Simon, one of Constantine's crew, replied nervously, "Simon Walters here, sir. He was there only a few minutes ago. I did feel a tug on the line. So, I looked back and thought I saw him and continued on. Wait—I just pulled up the tether line. The end of the line has his tether with the grommet still attached. Something must have torn it off his snowsuit. I'm so sorry, sir. Over."

---

23 See https://www.e-education.psu.edu/earth103/node/990   that documents the current extraction rate of 1 meter per hour and assuming 8 hours of work per day..

Holbrook came on with terse orders. "Constantine, take your crew and search the landscape. Follow our tracks carefully until you find Charley or his tracks. See if he diverged in a different direction. The rest of you, let's get these sleds into the craft."

Mitch expected Jake and Janey to object to being excluded from the search crew. Oddly, they were silent. The four of them started sliding the sleds up to the ramp, then entered the craft while the extraction team nestled the sleds with the cores into the holding space of the hovercraft. Mitch watched this last part of the procedure, wrote his observations, then returned to the team in the main part of the craft. After shedding their gear, they gathered in silence. Jake and Janey seemed more than worried, as if they already knew what was coming.

Two hours later, the expected nightmare became real. Constantine and his crew returned to the craft and delivered the bad news. Constantine said softly, "We found Charley at the bottom of the Reinhold Chasm. We found his helmet near the point where he diverged from us; that explains why he couldn't follow the GPS track, but it's still a mystery how his helmet came off. He must have hit the ice ledge squarely. He is dead."

"No! It can't be!" Janey ran to Jake as he put his arms around her. Her sobs broke the silence as the relentless wind-swept snow lashed at the hovercraft's windows.

Captain Holbrook said gently, with great affection and compassion, "I am so sorry for this loss, but I am especially sorry for you two in losing your cousin. My crew and I will take the sleds and retrieve the body before it gets buried in the snow."

Jake looked sullen, while Janey sobbed and said in a low quivering voice, "We'll break the news to his family." Holbrook joined Constantine and his crew and went out into the storm. For three hours, the silence was unbearable; finally they returned. Janey, beside herself, walked to Charley's body, kissed his frozen face and ran her fingers along his bearded cheek. She stood for a moment, looking at all the crew. "May he know that the great sacrifice he made today will bring us one step closer to solving the climate puzzle."

Everyone nodded. Captain Holbrook saluted, Jake and Janey's lips quivered, and Mitch lowered his head, feeling quite guilty at having poked fun, even if only in his mind, at Charley's red hair and laugh. Then Holbrook had the crew take Charley's body away.

Holbrook interrupted the silence. "Okay, people. Let's get these movable cores on the road, shall we?" It was actually a blessing to be doing something, and moving forward. Mitch had signed off on all of the remaining transfer procedures and their part of the mission was completed.

The evening was mournful, with only an occasional effort at light conversations to ease the burden of tragedy. It didn't help that Charley's banjo sat on the table where he had left it, staring at them all. Mitch was both sad and angry and took Captain Holbrook aside. "There is one thing I'm still curious about, Holbrook. I thought you said that you had all of the chasms mapped and that we were traveling on a safe route."

The captain looked irritated. "Look, Mitch, his tether came off and at some point he lost his helmet. Who knows where he wandered? Without the GPS map in his visor, he was blind. Charley probably wandered off the path into unmapped territory and fell into the chasm." Holbrook grabbed Dr. Mitch's forearm firmly and with a determined voice eased his anger. "Trust me, there will be an investigation into this tragedy. Now let's move on. Mitch, when we get back to the station, we'll be filming you overseeing the coring procedures and we will show the presentation over our Unity-vision broadcast. You're tasked with explaining to the public the coring procedure and the significance of interpreting climate conditions of the last million and a half years, so they can understand what our project is all about."

Mitch felt overwhelmed. "I can't do that!" he stammered.

Jake grabbed Mitch's shoulder in an effort to calm him. "Don't worry, it'll be easy! You just target your explanations to the high-school level, as the broadcast will be oriented to that age and up. Complexity is not helpful in a broadcast like this. Simplicity and clarity will make it understandable to the largest number of viewers. This is still a transponder-based Unity-vision broadcast; it allows people all over the world to instantaneously watch a screen that pops out in front of them, just like the one you jumped into for the debate twelve years ago."

Mitch was still uneasy. "How can I double-task this? I won't be able to observe the sectioning and sampling of each section of the core if I am being filmed!"

Holbrook reassured him, "Not to worry, Mitch. Jake and Janey will be working with the science team to do the quality-control observations. You have switched roles and become our documentary host."

Mitch then grasped the value of his two guides functioning as graduate students and realized he could do what was being asked. "I can do that," he said confidently.

Over the next week, Mitch reviewed the recent climate change reports and practiced his broadcast presentation. He began watching other Unity-vision broadcasts and saw how honest and transparent they were. They did not feature actors pretending to be scientists, as might have happened in the old YouTube videos of the past. This would be the real thing: unscripted scientists working animatedly and explaining their research and its value to the public.

Ten days later, Mitch was at the Vostock Station, where an international team of scientists was sectioning the core. But as a beam of light focused on him, the signal that the broadcast was beginning, he called time-out. "Hey!" he yelled. "Is this light going to distract the dissecting workers?"

"Not to worry," replied Andrej, one of the team's technicians. "We don't even see it. The light is all focused on you. Besides, this core is a 'dummy,' not one your team collected last week. If there is a break here, it won't matter."

"Oh," said Mitch reassured, "a prop."

The Polish technician looked at him quizzically. "Don't know what 'prop' is. Is that American slang for soda drink?"

Mitch patted Andrej on the back. "You're okay, my Polska friend. Whether you are joshing us here or not, you are one funny guy."

As the camera came into focus on the countdown, Mitch gathered his thoughts, breathing deeply. He was ready. "Good morning, this is Dr. Mitch Warner, a member of the U.S. team in this international effort. Our mission is to retrieve and analyze ice cores in order to reconstruct the climate of the last 1.5 million years. We are broadcasting from the Vostok Station, one hundred kilometers southeast of the Ross Ice Shelf in Antarctica, where we collected the extracted ice cores yesterday. Most unfortunately, on the journey back to our hovercraft, we lost one of our crew members, Charles Addison, who perished in a fall. We would like to dedicate this broadcast to him. He was a gifted and valued member of our team who will be greatly missed and always remembered for his good sense of humor and charisma. Charley, may you rest in peace." Mitch paused for a moment. He took another deep breath and continued his "climate info broadcast" (as it appeared at the bottom of the screen broadcasting to the world population).

"I'm here to explain why we journeyed into the Antarctic blizzard to obtain four cores of ice from the Ross Ice Shelf, some part of which may fall into the ocean soon. This study is part of the worldwide effort to learn as much as we can about whether human-caused climate change is warming the world or not. These ice cores—taken from a depth of almost two miles—can help tell us what our planet was like over a million and a half years ago!"

"A little background: Our earth's climate, throughout time, has been cyclical. We have had warm periods, as we are now, interspersed with ice ages. Normally, the evolution of these shifts takes place over 20,000, 40,000, 100,000-or 400,000 year cycles.[24] Because the earth's natural thermostat is affected by the amount of CO_2 and other heat-trapping gases in the atmosphere, their levels influence the temperatures of our global environments and are very important. When CO_2 levels rise, temperatures rise. When levels of CO_2 go down, temperatures go down. It is important to bear in mind that these cycles reflect a broad average over hundreds of thousands of years. They are not to be confused with weather, which varies from day to day. So, when we say that the temperature of the last two centuries has increased by one and a half degrees Celsius, we refer to a totally

---

24 See http://www.indiana.edu/~geol105/images/gaia_chapter_4/milankovitch.htm

different temperature scale from that used to report day-to-day weather. Now, let's turn our attention to these remarkable ice cores from Antarctica."

Mitch paused and the camera zoomed in remotely to technicians cutting sections of frozen ice from the great length of the core in the tube. "These scientists are taking samples of the ice from different locations in the core." The camera zoomed in to the ice section being cut by Andrej, then closer to the core that revealed the label A27. "We know the age of each section of the core through chemical analysis. The labels mark the different time periods during which the ice was laid down. It is like counting a tree's rings to determine its age, but here the cycle boundaries are marked by rings of dust and ash deposited during the dry season of each year.[12] Andrej, how old is this A27 section?"

Andrej looked at his clipboard. "20,000 years old."

Mitch continued, "These scientists are measuring CO_2 trapped in the pores of the ice laid down at various times during the past. In the core subsection of this sample, there are bubbles of air that contain CO_2 that was trapped in the core at the time the ice formed—in this case 20,000 years ago. The CO_2 froze into the ice during the time the glacier was forming. When we know the age of the ice, we slice a section of the core and take it to the laboratory for analysis so we can measure the CO_2 trapped in the ice bubble.[25] As a result, we know that CO_2 levels at that time were," Mitch turned to Andrej. "How much, Andrej?"

Andrej, who was a geochemist and coring technician, looked straight into the camera and said, "175 parts per million. This compares with our current atmosphere's levels of 450 parts per million."

Mitch continued, "Then we use chemical isotope analyses to determine the temperature of the air when this 20,000-year-old core section was laid down[26]

Mitch turned back to Andrej. "Can you give us an approximate average temperature at that time?"

The camera zoomed to Andrej. "Approximately 14 degrees Celsius. We will refine this number later with our models, but that is about 5 degrees Celsius (16 degrees Fahrenheit) lower than today, based on chemical analysis of oxygen isotopes."

Mitch closed the broadcast with what he believed were some intriguing and provocative thoughts. "Let's look at the big picture of how we answer the question of whether the earth is warming. Forty years ago, the scientific community was quite certain that the temperature increases we've experienced since the Industrial Revolution would continue and start to melt glaciers, cause more wildfires, increase and accelerate sea level rise, and lead to other impacts that precipitate adverse effect on society.

---

25 See http://www.antarcticglaciers.org/glaciers-and-climate/ice-cores/ice-core-basics/.
26 See http://www.antarcticglaciers.org/glaciers-and-climate/ice-cores/ice-core-basics/.

Then something odd occurred. Our temperatures started to decline, even though CO_2 concentrations in the atmosphere have steadily increased; now the temperatures are still decreasing. This seemingly contradictory information presents a puzzle. Were the models predicting the rise of temperatures wrong? Is this merely a temporary phenomenon in the warming slowdown? Or is there something else working in our climate system that we simply don't understand at this time? Finding a definitive answer to climate warming is analogous to finding the last pieces of an important puzzle in order to complete its shape, content, and color. What is happening in Antarctica confuses the challenge even further."

"As the global warming of the rest of the world has slowed down and reversed, we see a decrease in the rate of sea level rise and fewer heat waves. The numbers of droughts and wildfires have also decreased over the past twenty years. The minuscule amount of mitigation that society achieved since the early 2000s cannot explain the temperature decreases. But here in Antarctica we have rapid *warming*." Mitch was emphatic when he said this. "The big indicator of this is the imminent break-off, or *calving*, of part of the Ross Ice Shelf, which we expect will be shed in the next few months. Average annual temperatures on this cold continent are rising as fast as the world temperatures were in the early twentieth century. Could it be that the increase in temperatures of Antarctic is caused by its' relatively larger amount of oceans in the southern hemisphere[27]? Or could it be that the CO_2 and warmth that is stored in the Southern Ocean, which circulates around this ice continent, is being ejected into the atmosphere from the ocean and causing localized warmth here at Antarctica?"

The camera zoomed back into the ice core. Mitch concluded, "In the meantime, we are adding another piece to the puzzle with our reconstruction of the climate from the million-and-a-half-year-old sections of this ice core." Mitch concluded. "I am Dr. Mitch Warner reporting from Antarctica." The broadcast ended.

Afterward, the broadcast crew stopped for lunch with handshakes and congratulations to all involved with the broadcast. Mitch gave his personal view to the team gathered at the lunch table. "I still think that we are in a temporary hiatus and the release of CO_2 from the oceans will cast us back to a rise in temperatures."

Dr. Hans Bjorgen from Norway responded with a cautious, "We'll see." Surprisingly, they both laughed. Later, Mitch talked things over with Jake and Janey.

"So, Dr. Mitch, it seems the Unity-vision's viewing ratings were at 63 percent participation out of the world's population of 9.3 billion people. You are now known by billions all over the world."

---

27 A much larger portion of the Southern Hemisphere is covered by water compared to the Northern Hemisphere and water has a significantly greater heat capacity than land. See http://www.todayifoundout.com/index.php/2011/12/the-earth-is-hottest-when-it-is-furthest-from-the-sun-on-its-orbit-not-when-it-is-closest/

This staggering piece of information was more than Dr. Mitch could fathom and he asked a very simple question, "How did I do on the world stage?"

"You did very well, Dr. Mitch," Jake replied. "You simplified the information as best you could to the wide range of educational levels of the audience."

Janey added, "It was hard to balance the message between being too simplistic for those who are more educated in science, and being too technical for those who do not know the science. If you think about this, if you were to write a book about all of the Doors you've visited, you would face the same challenge. Although I thought that you were too much on the technical side, I agree with Jake. You did very well."

Jake stood up. "Let's proceed." Mitch and Janey also stood up. A new waviness opened in the atmosphere in front of them. They jumped in. . . .

2162

They landed in the next period of Door #4. It was 2162 in the same ethereal cloud as before, suspended above a classroom. Mitch's descendent, Dr. Leah-Ann Stuart, was in the room introducing one of her students. His heart skipped a beat, his eyes filled with tears. Dr. Stuart then sat down with a group of other professors with a student at the lectern, who was looking nervous. Mitch guessed that this was a thesis or doctoral defense. Dr. Stuart broke the ice. "Okay, Tracy, you have been working on your doctorate in climate anthropology for nearly four years now. Show us your stuff."

Tracy cleared her throat. "Welcome to the world of climate anthropology. I appreciate your attendance and the wide range of backgrounds that we have here this morning. Climate anthropology is a relatively new and fascinating field that had its beginnings a century and a half ago, amid the great debate of whether climate change was occurring due to human activity or whether it was just a natural cycle. Amid this debate, the field of climate anthropology came into play, examining the effects of climate change on culture through the ages. In fact, in 2015, Dr. Stuart's great, great . . . let's say great-times-six grandfather[28] taught a course in Climate Anthropology at the Miami Center for Environmental Research."

"My doctorate has examined the period of the last 150 years, a period of confusion. I have developed an analysis of this period centered on the evolution of science's standard of statistical certainty that transitions a theory into a law. . . ."

Tracy continued her presentation focused on the history of what transpired, centered on her table of events. She used a hologram of time events to project the history:

---

28 We assume that the difference between 2152 and 2030 (122 years) amounts to six generations.

Time Period	Events
2020–2030	World temperature decreases unpredictably.
2024–2034	World restudies basis of climate change to evaluate need for mitigation.
2050–2060	Temperatures increase dramatically; climate change emergency begins.
2060–2070	World turns to geo-engineering and strives to eliminate fossil fuels.
2070–2080	Negative impacts of geoengineering predicted by skeptics occur worldwide.
2090–2095	New theory of climate behavior emerges with 99 percent certainty.
2099	World achieves 100 percent emission cuts and elimination of fossil fuel power.
2105	Studies determine that skeptics of 2020 were right in calling for more information before mitigation and that climate activists of 2020 were also right in calling for elimination of fossil fuels.
2140–2162	Climate stabilizes to new equilibrium; human culture adapts.

Tracy elegantly worked in the evolution in the statistical certainty of evidence that raised the bar from 95 percent to 99 percent proof to make a theory into a law. She pointed to the positive consequences of buying time to gain a better understanding of how the climate system worked before taking radical action. "Through the eighty years of social and ecological turmoil we did, indeed, learn enough about the complexity of the climate system and how this saved trillions of dollars so that a time-tested, targeted approach balanced mitigation with better engineering." However, she also presented the negative consequences of waiting—the many lives and species that were lost and the disruption of ecosystems and human infrastructure that occurred when the real climate emergency hit us.

Tracy wrapped up, "So, based on our analysis of what was known in the early 2000s and the economic and social realities of what may have happened and what did happen, science has now raised the statistical bar from 95 percent to 99 percent certainty in order to evolve from a theory into a law. After a century and a half of intensive study, we are now 99 percent confident regarding how our climate system works. We have now accepted, in 2162, that climate is sensitive to carbon emissions, but we also see that it operates in different ways from what we thought in the early 2000s. Civilization evolves, adapts, and science leads the way to eventually address disruptions and emergency. I hold doubts of my own. Thinking about Dr. Stuart's great times six grandfather in his time, perhaps, in

2020, we should have mitigated our emissions by 80 percent reductions of CO_2. On the other hand, I also respect the fact that the inertia against fully mitigating in 2020 bought us extra time for cost-effective technologies to evolve. When we deployed these emission reductions in 2100 we were in concert with the new scientific law of how climate behaves. Perhaps waiting 80 years to mitigate was not such a bad thing. Yet the price: we lost so many lives and our ecosystems were radically altered causing a loss of species worldwide. In sum, we have been in a two-century climate war. Who was right and who won the war? I leave that for you to decide. You have the data from my analysis and now I invite your questions."

Ten minutes of questions followed. During this time, in the bubble suspended above the classroom, Mitch turned to Jake and Janey. "What the hell? How am I supposed to learn anything from this? Climate change and civilization's reaction to it operates so much differently here from how it did in the first three doors. This is frustrating and not very enlightening."

Jake responded, "First of all, we were not in the same future but are in a different dimension, a different scenario, a different reality. We are in Door #4, not in the reality of any of the previous doors. We are traveling to places that have a different set of circumstances, based on decisions that were made in the 2020s, the flawed certainty that we held in the mechanics of the climate system, and the vagaries of chaotic processes that define climate. What you learn from this, Doc, is that each decision that we, as a society, make at any point in time will define a set of outcomes for the future. Yet, the decisions and the actual outcomes are different from door to door. That is the whole point of these journeys—how decisions evolve into different realities as a consequence of world circumstances and the actions that we take or do not take . . . "

Janey added, "We don't want you to get overwhelmed by this truth, Dr. Mitch. It will become clearer to you as we travel onward."

Mitch shook his head seemingly resigned to accept all of this. *Okay. I will just have to Let it Be.*

The panel spontaneously applauded. Dr. Leah-Ann Stuart stood up, proud as a new mother showing off her newborn. "And now," she said, "by custom, we will ask Tracy to leave the room as we decide whether she has passed her bar into the doctorate club."

Tracy left feeling confident—yet somewhat nervous as she heard chatting and laughter behind the door. Nevertheless, she still looked at the possibility that her 99 percent certainty of passing her defense could be wrong and that she could be very disappointed. After a while, the door opened and she was invited back in. Dr. Stuart offered her an Erlenmeyer flask of champagne and faced the group. The faculty raised their own Erlenmeyer flasks. Dr. Stuart smiled. "A toast to Dr. Tracy Garnowski."

The Gray Door (#4): The Skeptics

The scene of the conference room faded. Mitch, Jake, and Janey were back in the vestibule. "Okay," said Mitch, "on to Door #5."

Janey smiled and interjected, "Whoa, partner! Let's talk about what you just experienced. Did you learn anything?"

Mitch felt impatient. "We have to talk about this now?" he asked as he glanced at Door #5.

"Now," said Jake. "We do not want to journey through these new doors of reality and leave you as disoriented as you felt after your journeys through Doors 1 through 3. So what did you learn?"

"Oh all right." Mitch sat down, took a deep breath, and scratched his gray beard. Then he remembered that he no longer had a beard in his new, clean-cut, youthful body. "Well, first of all, I am not so freaked out with confusion as I was in my journey through the first three doors. I let go. I Let it Be."

Janey smiled. "I think that you are actually having fun, Dr. Mitch."

"Perhaps so . . . but being calmer makes me more introspective and I am seeing the reality of my 2020 life, or dimension, or whatever it is, in a different light."

"How so?" asked Jake.

Dr. Mitch frowned. "It's hard. I am so certain about my views of the dangers of an impending climate change emergency. It is troublesome to see a dimension that questions my convictions. Nevertheless, I was able to accept that there are other perspectives to this debate. Even though I consciously accept the fact that science is never 100 percent sure, I feel certain that the climate change theory of my world is basically correct. Yet the journey through Door #4 has opened my consciousness to the infinitesimally small probability that I may not be right. It was enriching to experience this world."

Janey looked at Mitch with a dreamy stare. "So what do you think of this world that we have seen? I mean what do you make of the reality here, compared to the other three doors?"

"I know what you mean," responded Mitch. "I am seeing this as much more than a dream. That itself calms me down. A more expansive view of where we are on this journey through the doors is that we are in an alternate reality, perhaps an alternate universe. I am confident that I will return to my world of 2020 looking through new eyes . . . and that is arousing my curiosity to explore further in this 'long strange trip' we are in."

Jake and Janey laughed in unison. Mitch noticed that Jake's horse laugh was gone and he was less manic than he'd been when he first greeted Mitch. Perhaps Janey balanced his behavior.

"What's so funny?" Mitch asked.

Jake smiled. "That last line—'the long strange trip that this has been'—wasn't that from a group called the Grateful Dead back in your era?"

Mitch laughed. "Yes, I suppose you are right. The Beatles started us off, now we switched to the Dead. What next?"

Jake and Janey looked at each other and said in tandem, "You have learned a lot in this door, Dr. Mitch." Jake passed the baton to Janey.

"And now we must leave."

As they entered Door #5, Jake saw a vision of the green field that they were in before as they floated through Door #4 in that green meadow. He heard the Beatles, again. Now they were singing "Across the Universe."

Endnote, Chapter 4:

In today's world, the consensus of mainstream climate scientists is that CO_2 and other greenhouse gases are rapidly increasing and approaching a flashpoint that would bring us into a world with unprecedented climate changes and conditions that society's adaptation would be difficult. The International Panel on Climate Change (IPCC) is a worldwide organization of scientists that continually reviews all publications regarding climate change that were published in reputable peer-reviewed science journals.[29] They act as a "science jury" on climate change. They periodically issue a report of the science findings. Their 2007 report stated that statistically, we were 95 percent certain of the reality of climate change, caused by humans.[30] Then, in 2014, the IPCC report stated that "climate change is unequivocal[31] That statement is as close to a 100 percent certainty as you can get. Nevertheless, no reputable scientist can claim anything to be 100 percent certain. Therefore, the mainstream scientists still hold *some* doubt that their theories of climate change may not be correct. In fact the 95 percent vs. 99 percent certainty of whether a body of knowledge becomes a law is only part of the picture. In fact, once a theory has been repeatedly tested and independently verified to support it, we can call it a "scientific law." Scientific laws can, in fact change. This is based on the fact that we can never be 100 percent certain of any theory or law.[32] This is as it should be. This is the scientific method. So, for the public at large: should we not hold some doubts about whether we fully understand the climate system?

---

29 See https://en.wikipedia.org/wiki/Intergovernmental_Panel_on_Climate_Change.
30 See http://www.independent.co.uk/environment/climate-change/scientists-95-per-cent-certain-that-climate-change-is-man-made-8778806.html
31 See https://en.wikipedia.org/wiki/IPCC_Fifth_Assessment_Report.
32 See http://www.livescience.com/21457-what-is-a-law-in-science-definition-of-scientific-law.html.

CHAPTER 5

THE SILVER DOOR (#5) : GEOENGINEERING

How much can we depend upon technology to bring us out of a future climate crisis? Is there a solution ready to be deployed? If so, should we use it to save the world—even on a temporary basis—given the uncertainty of the side effects? Let us explore this option as we enter Door #5.

***

Mitch was back in the vestibule and found himself in the personal world of his own thoughts. *So many times I have been on the single-minded track when it comes to the* climate disaster. Jake, Janey, and even Casey have been so supportive of this crazy time-travel process. It is opening my eyes to realities that I never considered possible. I have experienced vast new dimensions. Yet, the one thing I know is that as much as I have learned, I miss my rather simple life and my family, especially my dear wife, Leah. I hope she is okay with all that has happened. Sometimes I sense that she is looking at me from above. Mitch came out of his reverie and faced the doors. He opened Door #5.

Mitch found himself on a watery slope. "Here we come, Dr. Mitch, right behind you," said Jake. Janey and Jake were on his tail.

"And we'll try not to bowl you over!" said Janey.

"Where am I now?" Mitch felt the wetness as water went splashing up into his face as he was gliding down a waterfall. The rocks underneath were smooth and slippery. "Oh, my gosh, here comes the end of this slide!" The large pool at the bottom was coming up fast. In an instant, he plunged down into crystal clear turquoise water with a silver hue and emerged with Jake and Janey laughing out loud, while he came up sputtering—but it didn't take long for Mitch to start laughing also.

Mitch then saw the eye-opening slide that he was on—a watercourse that had a perfectly smooth bed of rock that allowed him to glide on his back, reminding him of a summer swimming pool plunge.

The three of them got up to the shore where they were faced with a silver door. It looked like an industrial-sized refrigerator door. "I guess we are entering a cold era," Mitch quipped. Mitch started opening the silver door.

"Wait!" said Janey.

Mitch paused, then asked, "What gives?"

Janey replied, "Are you ready for this next journey? We are switching to another set of gears. Take a deep breath and open your mind. For you, this will be another mind-blower and we are trying to guide you slowly through these alternate realities so that you are not so disoriented when you return to your home dimension of 2020."

After a pause, Jake said, "Now take another deep breath and follow us as we dive to the bottom of the pool. There is a silver door there, which is the real one. What you started opening was the door of a refrigerator. Follow us into the door."

Mitch wondered what was next. *Well, I sure don't know what to expect, but after the skeptics of door #4, I am ready for anything. Here I go.* He dove into the water and emerged in a silver-lined cavern. Jake and Janey were already in there, facing a silver door—the real silver door.

Mitch's thoughts were still in a state of curiosity as he entered another world. *Admittedly, these quests are getting more and more interesting.* He followed Jake and Janey through the silver door.

2032

As he opened the door, he faced a world where there was a silver tint to everything. Yes, it was the year 2032. Same time, different channel. He found himself in a laboratory with workers at various stations looking into surreal hologram screens in front of them that just hung in the air. One of the workers who seemed to be the lead man was going from station to station checking what each person was doing. He turned as he spotted his guests. "Jake, Janey, I have been expecting you." He walked over to them and extended his hand out to Mitch. "Hello, I am Dr. Henry Pazderski, at your service. Come, let's sit down." The three of them sat at a conference table with Dr. Pazderski. In front of them were the same square translucent boxes they had encountered in the 2032 world of Door #5. "Do you want something to eat?"

Dr. Mitch smiled. Pazderski was somewhat tall, thin, bald, and with an energy that seemed ready to brim over. "Well, a slice of pizza and a beer wouldn't be bad." After Pazderski waved his hand in front of a silver door, the food cooked in one of the boxes and slid open to present Mitch his meal.

Mitch opened the discussion. "So, what year are we in? Let me guess, 2032."

"You catch on fast, Mitch," Pazderski responded. "Now let me cut to the chase and give you a briefing on what this dimension is all about."

He told Mitch about this world reality, with Jake and Janey filling in some details. The world had entered a rough period in the 2020s as the full wave of climate change impacts rode over the planet. A good part of Bangladesh was gone. Coastlines were flooded worldwide and the world economy tanked. Tensions between nations erupted into eco-wars over scarce resources, mainly water. New disease epidemics appeared year after year, due to new vectors that resulted from a warmer climate—particularly in the tropics.

When Pazderski paused, Mitch pitched his first reaction. "This seems very similar to the first three doors that I entered."

Pazderski looked confused. "First three doors? I don't know about that."

Jake whispered to Mitch, "Let it Be, Doc. One reality does not really know about the other dimensions that you have visited. We don't want to confuse the inhabitants of this world with the same confusion that you had when you returned to your 2020 world. He only knows that your skipping stone delivered you from a different time and place. He doesn't really need or want to learn more."

"Never mind, Pazderski," said Mitch. "Please continue."

Pazderski resumed his briefing. "National policy everywhere on the planet was eclipsed by international policy and a world union spawned out of the U.N. entitled 'Save the Planet.' Quite original, huh?"

Mitch laughed. "At least it wasn't one of those long acronym-enriched bowls of consonants. And what other name could showcase this international emergency?"

Pazderski then described a global crash research program (similar to the work to land on the moon) to find an immediate solution to the crisis. Geo-engineering was the only quick fix for the climate emergency. "Are you familiar with geo-engineering?" he asked.

"Oh, yeah," said Mitch. "I don't want to get on my soap box on that and will need to try extra hard to keep an open mind. So let me guess. Your organization—whatever it is—is working as part of this international initiative to model the potential geo-engineering outcomes. Am I right?"

Jake and Janey clapped in unison. "Bingo!" said Pazderski. "You are very perceptive . . . you mind if I continue calling you Mitch?"

Mitch smiled and slapped his shoulder. "As long as you don't mind me calling you Pazderski."

"Okay. On with our briefing. I want to tell you what your role will be."

Janey cut in with a whisper to Jake and Mitch. "Your mission, should you choose to accept it, Dr. Mitch . . . " Mitch laughed while Pazderski looked puzzled.

Jake continued ". . . is to work with the team as their ecological skeptic."

Janey chimed in again. "This recording will self-destruct in one minute."

Mitch laughed and whispered to the twins, "*Mission Impossible*, from my dimension, as you call it."

Pazderski looked up. "Are you three finished whispering your jollies?" They all nodded and he filled in the details of how Mitch would interact with each of the technicians at their work tables to create future scenarios that would accurately reflect the ecological impacts of what the world would face with deployment of the selected technology. "So let's go to the exhibit hall next door. It graphically depicts the four main geo-engineering options."

They walked through a door labeled "Displays of Geo-engineering." Pazderski walked them toward the first display. "Mitch, I know you will be surprised at the concepts, if not skeptical. The explanations are geared toward the non-technical general public but I can answer any of your more engineering-based questions. Nevertheless, you will appreciate the visuals."

The first display was entitled "Seeding the Oceans." Instead of a sign or a kiosk, the explanations were shown on those translucent air holograms suspended in the air above each display.

In the wavy air above them, a scientist, Dr. Maureen Shelby, came into view, introducing herself as an oceanographer and then directing the audience to the model behind her. She explained the scheme of dumping iron from ships into the Southern Ocean. She described how microscopic plants (phytoplankton) are limited by low iron and will not grow past a given population level until there is more iron in the water. "So if we discharge large quantities of iron filings into the ocean surrounding Antarctica, the phytoplankton population will explode. This large bloom will have an effect on the atmosphere. Since phytoplankton inhales CO_2 out of the atmosphere, the concentration of this greenhouse gas would be reduced in the atmosphere and temperatures on Earth would decrease."[33]

The display of the iron-dumping scheme had three separate components: The first module showed the existing condition of the area around Antarctica (prior to injecting the iron filings). The module showed the continent and the sea water of the Southern Ocean. There were small amounts of phytoplankton floating around Antarctica. The concentration of atmospheric CO_2 (460 parts per million) was displayed on a readout on an ethereal screen above a model of the area around Antarctica. The second module showed the same area with a fleet of ships unloading iron filings into the ocean. Above the second display there was a sign that read "50 years after project inception" and a readout of the CO_2 in the air above the module that showed 400 parts per million of the gas. Dr. Shelby explained that the representation of the phytoplankton in the sea showed

---

33 See http://voices.nationalgeographic.com/2012/10/18/iron-fertilization-savior-to-climate-change-or-ocean-dumping/.

approximately ten times the amount of phytoplankton as the first module. The final display was labeled as the "full operational stage" of the project, projected to be 100 years into the future. It showed the same ships dumping the iron filings, but this time the sea around Antarctica was fully over-bloomed with phytoplankton and the display of the CO_2 concentrations showed 325 parts per million, which was what the levels were before the Industrial Revolution in the 1800s. The projected average annual temperature of the atmosphere was 55.8 degrees Fahrenheit, approximately the earth's thermostat reading in the early 1800s.

At the exit point for this display, Dr. Shelby appeared again and explained that the iron-dumping technique had been widely discounted due to the uncertainty of how it would affect the dynamics of the climate system, the ocean, and the limited ability of "pulling the plug on" or terminating the process, if the environmental effects proved to be out of synch with the models and exhibited negative impacts on the planet.

Mitch said that he was impressed with the display but commented that based on what he knew about geo-engineering, the same conclusion should probably be announced after each of the subsequent rooms.

"Now keep your mind open and Let it Be, Dr. Mitch," commented Jake.

"Okay, okay," Mitch responded and they entered the second room. The sign above the room said "Carbon Capture: A new technology.[34] There was a model of a large chamber with a suction fan taking air into it and exhausting it back into a hole in the floor. The casing for the fan was glass to show the material that the air flowed through. The media in the chamber was composed of fused silica fused with a polymer, the sign explained. It appeared as a pink foamy substance in the glass representation of the inside of the chamber. Dr. Carl Bertrand appeared on the hologram above the display and explained that this model (which took up most of the room) was only a scaled-down representation of what the area of the chamber would look like. The real chamber would occupy two miles per station. Dr. Bertrand explained to them that the air flowing through the chamber would be scrubbed by this foam and that this would take out most of the CO_2.

The next display in this room illustrated a global map exhibiting locations of all of the stations where this chamber/fan technology would need to be deployed if it were to be effective in lowering CO_2 in the atmosphere. The pink points on the map were all over the land surface, one million in all. Just beyond the display, Dr. Bertrand appeared in a screen above them again and stated the advantages and drawbacks of this technology. "The advantage of carbon capture is that it represents a large-scale emissions control project of the ambient air, offsetting the amount of CO_2 that we inject into the atmosphere from

---

34 See http://energy.gov/fe/science-innovation/carbon-capture-and-storage-research/overview-carbon--storage-research.   NOTE:This technology is in its infant stages with a single protype plant. The data presented below (regarding the number of plants required and their costs) are entirley fictional.

power plants, motor vehicles and industries. The precise number of fans would be determined by the need to offset 100 percent of our emissions of CO_2. If the amount that is removed proves to be significantly greater than that which is emitted, it could be adjusted simply by turning off some of the fans. The disadvantages are the high costs for constructing and installing the fans and the large areas of the earth's surface that would be needed for the approximately one million that would have to be deployed, with each site needing twenty-five to fifty acres. It also displayed the approximation of the costs of each station—$57 trillion—along with a $1 trillion per year operating and maintenance budget."

As they exited the room, Dr. Bertrand appeared in the hologram one last time. He explained that carbon capture was a technology of the future and that a deployment of this prototype would simply require a final engineering and site acquisition process that might take twenty-five to fifty years. He also stated that this technology could be used to replace the space-based technology exhibited in the next two rooms. "It is a grand technology, but much too expensive," he said. "Maybe someday research will make this scheme practical."

As they exited the room, Mitch again commented on the display. "Fascinating--though we did know about this technology back in my dimension. But seeing it on a scale model as represented by working prototypes is certainly an advance. Seeing the vast numbers of these fans needed and the costs is an eye opener. Don't worry, my mind is open."

Pazderski smiled and took them through the third room of the exhibit. "Now we go to the space-based options, which are closest to actual deployment." The room had four scale-model prototypes of spacecraft suspended above them, and a fifth one on the floor in the center of the room. Each ship was discharging a fine orange mist that remained suspended in the air, slowly falling down. Dr. Tom Harley appeared on a hologram screen above them and explained the engineering concepts. "A direct way to control the earth's temperature is to reduce the incoming solar radiation One way to do this is to send a fleet of spacecraft into the stratosphere and discharge sulphate aerosols.[35] As you can see here, the aerosols remain suspended in the air, like a cloud, though they slowly descend to Earth. These aerosols reflect incoming sunlight and therefore more directly reduce the average climate temperature." "Wow," said Mitch. "There is still serious thinking about that type of geo-engineering.

Pazderski turned to him. "Wait, you haven't heard the last explanation for this room."

Dr. Harley appeared again. "Although this type of geo-engineering was, for a while, a leading candidate for controlling Earth's temperature, detailed studies revealed many drawbacks, despite its lower costs. Sulphate aerosols eventually do make their way down through the atmosphere where they will precipitate out as acid rain. Despite an effort to find a way of ameliorating the acid rain effects by injecting buffering agents into the

---

35 See https://en.wikipedia.org/wiki/Stratospheric_sulfate_aerosols_(geoengineering).

aerosol cloud, it is generally thought that there is too much uncertainty about whether we could avoid the acid rain. Acid rain from industrial emissions plagued the earth in the late twentieth century until nations instituted emission controls for industries. This now leads us to the final alternative for controlling Earth's temperature by engineering. Please proceed into the next room."

Mitch deadpanned, "I can't wait."

The fourth room of the geo-engineering exhibits had an eerie silver air. Not translucent—you could through these silver shimmering objects. It was dominated by a spacecraft that filled the room and four others above and adjacent to them, just as in the third room.

"I figured this would be the gem of the show," quipped Mitch.

Calvin Curtiss appeared in a hologram screen above them. "The five-year study that was concluded in 2031 is likened to the Manhattan Project that created the first atomic bomb in the 1940s. In the 2020s, the onslaught of the first wave of the climate emergency prompted an international research effort led by thirty-one nations to determine the safest and quickest method for us to put a stop to the warming of the earth. The initiative was named the 'Solar Reflector Solution,' but the media and the public quickly renamed it with a less technical term: 'The Silver Spaceship Solution.' In a mere three years, 'The Silver Spaceship Solution emerged as the leading alternative for controlling Earth's temperature. At that point, the initiative evolved into a full engineering development of the spacecraft fleet. It is now ready for deployment. In front of you is a one to one hundred scale model of one of the spacecraft in the proposed space-based fleet of sun reflectors. The mechanics of its operation are intuitive, as you will see."

The four of them stood in silent awe of the ship that had multiple arrays of mirrors and silver surfaces. Each ship was discharging a fine mist of tiny silver particles—nanoparticles[36]—that overlapped the silver cloud emitted from the adjacent ship. This is what gave the air in the room a silver sheen. It was clear that the operation of each ship was simply to emit a silver cloud that reflected incoming sunlight back into space.

Mitch broke the silence. "Back where I came from, this technology was considered a joke, though some research was done to study its feasibility. The policy was to have such a solution in our back pockets in case a climate disaster descended upon us with no time to reduce our emissions. I, however—"

"Hold your thoughts, Mitch," said Pazderksi, "until you hear the closing summaries."

They came to the next monitor. Calvin Curtiss came on the screen. "After the three-year alternative analysis was completed, prototypes were developed among the world nations leading the study. The next phase, which we entered in 2032, compares the different prototypes and the best design for the fleets. At the same time, we will be conducting extensive modeling of the temperature reduction and possible negative environmental

---

36   See patent application, http://www.google.com/patents/US20100127224.

impacts. It was decided to centralize this latter task at one location: here at the National Center for Atmospheric Research (NCAR). And yes, there is still opposition to this technology. Please proceed to the next monitor."

The group stopped in front of the next tour location. A woman in a yellow bodysuit appeared. "Greetings. I am Valerie Kelly, President Rother's EPA Administrator. Many individuals and organizations are still questioning the wisdom of proceeding with the launch of the so-called 'Silver Spaceship' fleet, as it has been dubbed. Even though we are in the midst of the climate emergency, EPA has been charged with the responsibility for evaluating planetary side effects. To focus on the objectors, we are, indeed, manipulating the earth's thermostat. Yet, we have our backs to the wall. EPA is addressing this problem with a two-pronged approach. First, we are conducting a critical evaluation of the modeling of the effects of these reflectors by an interdisciplinary team. This group includes an environmental engineer, a hydrologist, a geophysicist, and a climate ecologist."

Photos of the selected scientists flashed in front of them briefly. One of the photos caught Mitch's attention. *It can't be. I must have imagined that. That last team member couldn't have been who I thought it was.*

The EPA Administrator continued. "This team will be working with the climate modelers at NCAR. The team that has been selected is considered the best and the brightest. They will *not* be doing new research or modeling. Instead, they will be serving as a 'science jury' of evaluators. They will extensively evaluate the effects of lowering the solar radiation input to the Earth from the sun. The team has been assured that they will have a significant input to the world's top decision-makers regarding the launch of this technology. And this will happen before the two-year deadline for the proposed launch of the so-called 'Silver Spaceship' fleet. The second initiative is to create a structure for a massive program of emissions cuts to reduce our CO_2 levels to pre-industrial revolution levels. This will assure us that we will be deploying this technology as a temporary Band-Aid and wean ourselves off of it to transition to a green age of living within the limits of our planet's carrying capacity. Please proceed to the next station."

The final hologram screen concluded the tour. The President of the United States, Julie Rother, appeared in front of them. "Good day. You have just witnessed the status of world geo-engineering. We hope that you have learned about our challenge and we invite your input through your Unity vision transponders. We will be listening to your critique. However, this tour has also been designed as an introduction to our team of evaluators who will hold center stage over the next two years. This team comes from a variety of backgrounds, life experiences, and skills. It is largely their input that will determine whether we launch or not, and if we do, what restrictions we must implement. However, as Administrator Kelly has said, our backs are against the wall. Every day we are losing more of our coastlines to inundation by the sea. Every day, the environmental impacts are wreaking havoc on the earth. You may ask why we are putting all our faith into a team of four individuals instead of deploying teams of scientists from around the world. We do this to cut through the bureaucracy. In no way could multiple teams complete an assessment in two years. So, we selected the best and the brightest based not only on their scientific qualifications, but also based on their integrity and independence. Although they will have access to ecologists, engineers, geoscientists, and experts in other disciplines, we have numerous safeguards in place to assure that they will not be pressured by lobbyists, bribes, or by their own governments. In fact, there will not be any limits to their funding nor will there be any personal contact between them and policy makers. The team is not charged with research. They will function as a team of jurors to decide on the evidence that has been collected over the past ten years. Two years is a tight but reasonable deadline for them to evaluate the technology and report their findings."

President Rother drank a sip of water and continued. "Efficiency, brains, and fierce independence is what we need to make the final decision by October 1, 2034. And as we speak, three nations—China, Japan, and Brazil—are on the threshold of developing their own independent fleet to 'go it alone' with this engineering. The U.N. has avoided gridlock on the question of stopping the individual nations' deployment of the 'Silver Spaceships' with a promise that we will have a decision on whether to proceed internationally by 2034. In exchange, we are assured that they will participate with us—with open minds—on the ecological evaluations and the modeling of the deployment of the fleet. It took us ten years, to engineer and launch a flight to put a man on the moon in the 1960s. We need to complete the current project in two years—and do it in a manner that will not be a mere rubber stamp. I have confidence in our team. I ask for Americans and world citizens to stand by us in this moment of crisis. God bless our fragile planet. Thank you."

"Okay, guys and gals, let's go back to the conference room to discuss Mitch's role," said Pazderski. They followed him to the conference room.

Mitch looked at his two compatriots. "What's the matter, cat got your tongues? You two were really muted out during the tour."

Jake replied, "Hold on tight, Dr. Mitch, keep your mind loose and open, you are about to learn something very interesting about yourself."

Mitch looked at his two guides quizzically. An old twentieth-century tune rolled into his head: *"What Will Be, Will Be."*

Janey seemed to read his thoughts in her reply: **"*Que Sera, Sera ... and Let it Be*"**

They assembled in the conference room. Pazderski offered them coffee. "No thanks. My nerves are too much on edge already. I am anticipating something here," said Mitch.

Pazderski sat down. "Okay, let's get to it." He leaned toward his office intercom, "Margie, send in the team." A few minutes later, three people entered the conference room. They looked vaguely familiar to Mitch. Pazderski began his introductions. "Mitch, these are our teammates. Meet Dr. Mya Che-Ling, our geophysicist." Che-Ling, a young Asian woman with long dark hair and a sweet smile, shook Mitch's hand. "And this is Dr. Simon Hanaford, an environmental engineer from Saudi Arabia." He was middle-aged, looked very fit and very bald. He approached Mitch and shook his hand. "Finally, meet Dr. Tom Harley, our hydrologist. Harley was a bit older than Hanaford, sported a trim mustache, and had a quaint English accent that begged to charm. He faced Mitch and bowed, "Pleased to meet you."

Suddenly, Mitch knew why they looked familiar. "Wait, you guys and gals are the team selected to be the review team—oh, let's call it a jury—for the 'Silver Spacecraft' launch! I recognize you from the images in the last module room of the exhibit. It's an honor to meet three of the four of you. I hope that I will have the honor of asking you questions and providing my input."

Everyone in the room laughed, save Mitch. He looked alarmed as something started to dawn on him. "There were supposed to be four on the team. Is the fourth—the climate scientist—going to join us?" Again the others laughed.

Pazderski answered the question. "Mitch, I think that you know where this is going. You are the fourth."

Oh, my God, I guess I didn't quite accept the identity of the picture of the fourth team member because it is the younger version of me that I am not used to. How is this possible? How could this reality know of my qualifications and experience and why would the president pick me? Two years? We are supposed to analyze and make a major science policy recommendation in two frigging years? Why me?

Mitch vocalized his angst, boiling it down to the "why me" question that was floating in his mind. "I don't see how I can make this team work with the super-efficiency that the president wants. You have got to know, back where I came from I was an extreme critic of the geo-engineering technology that we now seem ready to launch. I will be the rebel that you three will think is the devil by the end of two years."

Dr. Che-Ling responded. "Mitch, we are all rebels—or skeptics if that fits better—which is precisely why we were selected. We all accepted our posts and are depending upon you to join us."

Dr. Hanaford chimed in. "We need you for your fierce independence as much as your knowledge of how climate change affects the dynamics of ecosystems. I read your papers in the science journals. We need you."

Dr. Harley added, "Look, I know that this is disorienting, but the three of us studied your personality profile as much as your science record. Of all the climate ecologists out there, we felt that you would be the best fit. Our goal is incredibly difficult and we need to work like a synchronous and efficient machine to meet the October 1, 2034 deadline."

Mitch was, indeed, disoriented and a little irritated. "Well, I don't know how you were able to get my personality profile, but what makes you think that I will be the perfect team player?"

They all laughed. Dr. Che Ling responded, "That is exactly the profile we seek. We know that you have strong convictions. We know that you have a bit of a temper, but your record shows so many accomplishments working with teams of four or five people. In a larger or smaller group, you would not have fit the team profile."

Dr. Che-Ling added, "We have been given not only assurances of our independence but our orders are to analyze with an unusual default position: that this geo-engineering proposal is too uncertain and risky to launch. It is up to the rest of mainstream science—and their models—to prove that our skepticism is wrong."

Mitch continued, "It's true that I am very skeptical of the scope of our work. How in God's name can we research all of the intricacies of knowing the time-tested thermostat of our planet and recreating it with a thermostat of our own through geo-engineering . . . in two measly years? This seems impossible!"

Hanaford responded, "Mitch, Mitch, all of that work has been done for us. We will not be re-inventing the wheel after ten years of research on the feasibility of the Silver Spacecraft fleet and all of its effects on climate and the environment. It's all there for you to examine. And that is the key word. We are the 'assessors'—the jury—that is doing an independent evaluation of all of the modeling data that is available to us. Our charge *is* to try to shoot holes through the current consensus that the spacecraft fleet will work in lowering the earth's temperature and that any environmental side effects, if there are any, can be handled by adjusting the amount of sunlight coming in to the planet."

Mitch looked unconvinced. "They say that they are not pressuring us, but the weight of our mission—being the last hurdle before the international Silver Spacecraft are launched—seems pressure enough. How do we know that this is not a ruse and that we will just cave in and merely give this scheme a rubber stamp of approval? After all, the world leaders are in a panic to solve this crisis quickly and impulsively."

"We don't know for sure that they haven't put us in the position that you describe, Mitch," said Dr. Che-Ling. "But I take them at their word that they will leave us alone as we do our work. If they do not, I will resign. But as long as they keep their word, the onus will be on me—and us—to be diligent in performing our evaluation in a vacuum, blocking out the weight of society's pressure for a quick fix to our climate emergency."

Mitch considered this. "I appreciate your honesty, Dr. Che-Ling."

So, in the end, Mitch agreed and joined the team. Pazderski summarized their tasks: "The analysis of the models that some of you have already begun is only part of your charge. I expect you to work separate from each other on this, but don't be afraid to periodically meet to share notes. The main objective of this first phase is for each of you to draw your own independent conclusions on the feasibility of deploying the Silver Spacecraft fleet in two years. When you are ready—keeping in the back of your minds the October 2034 deadline—you need to become an interdisciplinary think tank that will act as a jury for the proposal. No one will influence you on this. The U.N. Council on Geo-engineering Implementation will meet with you only after you have submitted your deliberation package, to frame how they will announce their decision about whether they will deploy or not. If you have a 'hung jury' scenario, you need to contact me and we will have to deal with that, respecting the dissenter or dissenters. Remember—you will not be alone. You may draw from the resources of the technical support specialists in the scientific communities around the world. You only need to document each outside discussion and be prepared to shut down your contact if you start receiving pressure on your decision. This will be a tough job and I bid you well. Now let's get to work."

All of them left the conference room and re-entered the NCAR laboratory and started immersing themselves in the models that were being run at thirty-four different stations. The final evaluation process had begun.

March 1, 2034:

Since 2032, the team members independently gathered their information. Mitch thought about this. *Wow. A year and a half without Leah and Susannah. But wait, this is all happening in one night! I don't want to think about this, as it will just throw me into a loop of confusion.*

By March 2034, they were ready to deliberate. The four team members convened in the conference room. All four had completed their analysis of the feasibility of a launch of the Silver Spacecraft fleet and the environmental effects. All of them realized with some concern that they had left very little time to come to a decision.

Pazderski entered the room. He was followed by a tall black man with long dark wavy hair that was tied into a ponytail. He was wearing a leather vest, a string tie, a wide-brimmed Indiana Jones hat, and clean-cut dress jeans. He also wore a minister's collar. On anyone else, this outfit would look ridiculous, but somehow it fit this man. It was hard to pin down his age. He seemed on in years but would be best described as seventy-five going on sixty. He seemed to have an aura of utter calm. Pazderski opened the discussion. "Team, I want you to meet Reverend Dr. Richard Tyndall. He is a graduate of the Yale Divinity School and received his doctorate in 2018 in environmental ethics. He will serve as your foreman. As I told you two years ago, I am not part of the jury and I will intervene only if it appears that you are headed for a hung jury."

Reverend Tyndall joined the group with introductions around the table. "Greetings, everyone. I understand that all of you—except Dr. Warner—prefer to be addressed by your last names: Hanaford, Che-Ling, and Harley. For whatever reason, Dr. Warner chooses to be the non-conformist and is just Mitch." Everyone in the room laughed. "I also understand that Mitch has been silently shadowed by Jake and Janey Smith, as some sort of chaperone from the place Mitch has come from."

They both smiled. "Ask us no questions," said Jake.

". . . and we'll tell you no lies," added Janey.

"Yes, I have been briefed," said Tyndall. "No one talks about where Mitch has come from, we just accept it. As for me, we'll keep it informal. Call me 'The Rev' ". Everyone laughed again.

Pazderski broke in, "As most of you know, Reverend Tyn . . . Okay, The Rev . . . is President Rother's science advisor for environmental ethics. His expertise lies in environmental mediations. He is well-equipped for this job. Significantly, he will be a blank slate regarding his perspective on geo-engineering and has no vote on your jury. I don't know where he stands on this geo-engineering issue. His role here is to simply run your panel's discussions and keep things moving along, and guiding you as to where he thinks the consensus is shifting, one way or another. He will, of course, guide the order of your process, but I think that you will find him very easy going as he does this. So, enough from me. I exit the scene at this point and leave you to your first interdisciplinary discussion for the decision that you need to make. One last thing: You all know the October first deadline for your decision and that is the last you will hear of it. Proceed as if you have all of the time in the world. This may sound counterintuitive, but The Rev will be guiding you toward that decision without pressuring you to make it soon. He will get you there. Good luck to all of you. We will talk again after you have made your decision." Pazderski left the room.

The Rev smiled and said to the group, "Hey, how about those Cubbies? They finally learned how to put together a team to win a national league championship!" Everyone laughed. Mitch was confused as he thought that the Chicago Cubs won the NL Championship in the late teens. *Let it Be*, he thought. *I am in a different dimension now.* A period of levity followed. The Rev knew it was time to start. "Okay," he said, "ground rules. I will guide your flow of ideas and viewpoints to keep order. I will not offer any of my views or influence your deliberations in any way. When things get hot I will ring this chime . . ." The Rev took a single copper tube out of his briefcase and tapped it with a small hammer that caused a long reverberating and very calming sound. "This means we stop, take a break for a few minutes, and continue our discussions afterward. Our meetings will be limited to two-hour sessions, once in the morning, once in the afternoon.

In between meetings, I encourage you to discuss the matters among yourselves and to recheck your notes or even revisit the people running the models if needed. I will also periodically give you my read on where the discussion is going. That's pretty much it, as far as I can see. Any questions?" Everyone shook their heads.

"Okay, let's begin! Let's start with a poll of how each of you currently views the viability of the Silver Spacecraft proposal after two years of your own research. Who wants to start?"

Hanaford spoke first. "I've spent a lot of time trying to poke holes in the technology. The engineering is sound and the Silver Spacecraft fleet will work in reducing our planet's temperature. Yet, I still have reservations on the financial investments needed to operate and maintain the fleet for an undefined period of time. I also recognize the possibility of unknown side effects on the environment, but we always have the option of pulling the plug on the program. I have many details to share with you on my analysis. But overall, I conclude that it can work. Deploying the fleet will lower the earth's temperature within five years and it may be less. From there, monitoring will be used to fine-tune the reflection, keyed to what is happening on Earth."

Harley was next, adding his perspective. "Well, as you know, I was looking at the hydrologic cycle. I examined the agreeability of the models and they corroborated at a 95 percent confidence level. The conclusions, however, are vague. Yes, blocking incoming sunlight will immediately affect the water cycle budget of the earth.[37] But from there, the climate system can go in many different directions. Some areas will have more frequent storms, including hurricanes. Some regions of the world will not be affected significantly. Others will experience less rainfall—the obvious overarching concern is whether the monsoon season will break down and rob Asia of the rainy season that supports agriculture. What disturbs me is that the vagueness of these forecasts suggests that we still don't know enough—and we will probably never know enough—about the hydrologic cycle to accurately predict how the earth will react. On the other hand, this is a climate emergency and every day, we are seeing more of our population suffer and die and sea levels will continue to inundate coastal cities. Yes, it is a gamble. But I yield to Hanaford's confidence that adjustments can be made with the reflectors to reduce any negative effects that we cannot predict at this time. I would say that it would be judicious to move forward with the launch."

Dr. Che-Ling seemed to be holding herself back from interrupting Harley during his summary and quickly chimed in. "As you know, I looked at the project from two perspectives. As a geophysicist, I am looking at the project's effects on the short- and long-term climate processes and how they relate to incoming sunlight. On that score, I fail the technology. We cannot ever predict how the earth system will react with any accuracy. Many

---

37 See http://www2.ucar.edu/atmosnews/news/10531/geoengineering-climate-could-reduce-vital-rains.

of the effects are subtle and we will not detect them for years to come and at that time we will not know whether it is appropriate to increase or diminish the incoming sunlight. My second perspective, as a systems analyst, makes me reflect on Hanaford's conclusion. The system will work and will provide an immediate relief from the onslaught of the climate emergency. I see one solution that could be our ultimate compromise. If we recommend deployment, we stipulate the recommendation with a condition that the fleet will operate for a very specific limited time—say twenty years—after which time the fleet will be deactivated to see where we are and to re-evaluate what is going on. Twenty years of deflecting sunlight will not harm the planet in any irreversible way."

Finally, Mitch had his say, and was clearly chomping at the bit. "I don't know how we could naively give them the green light with the hope that they will stop the system once deployed!" He found his temper rising and his face started to turn crimson red.

Janey softly stroked Mitch's arm and whispered, "Let it Be."

He took a deep breath and calmed down. "Okay. It is possible that the U.N. would take the spacecraft down after twenty years, but I think that it is highly improbable. As for my evaluation, I would echo both Harley's comments and Che-Ling's views on uncertainty. The earth's ecological systems will be affected, and in ways that we cannot predict with any accuracy. Look at all of the confident predictions of the early century that the temperature rise would be gradual and its related effects on the environment would be subtle and creeping up on us by mid- to late-century. Then, WHAM! The 2020s climate emergency came on us like gangbusters. The models were wrong. As much as I respect the integrity of the scientists who developed the models on the spacecraft reflector project, my own analysis showed anywhere from three to ten different outcomes, if one or more of the variables were under-predicting or over-predicting the effects. Where I strongly disagree with everyone in this room is the naive reliance on the ability of the Silver Spacecraft's reflectors to be adjusted to tweak the incoming sunlight. Che-Ling has it right with her concern that we cannot know in real time what the undetectable effects are out there and how to reflect more or less light. These effects can turn the earth's ecological systems into cycles that we may not recognize for decades and that we may not be able to reverse. I appreciate her suggestion of a compromise solution to turn off the reflectors in twenty years. But think about it, folks. If we give our green light to the 'easy fix,' will they ever turn back?" Mitch's voice started rising in pitch and his face twitched with the onset of anger. "This manipulation of the earth's thermostat is a Band-Aid and will forever deter us from what we really need to do, which is cut our emissions by 80 percent. Come on, folks, wake up and smell the marsh gas!"

Jake came over to Mitch, patted him on the shoulder, and whispered, "Think of what you learned from Door Four: the Skeptics."

Mitch took another deep breath, stretched his arms, and concluded. "I know that I have strong views here, but I must also advise you that I could be wrong. But I sure feel very confident about what I do know. I feel strongly that we need to put the brakes on the momentum to launch. At this time I would vote a definite no."

The Rev rang his time-out chime and came back into the discussion almost imperceptibly and in a calm, quiet voice, which was nonetheless commanding. "People, we have been at this for forty-five minutes and I can see that the discussion is starting to get hot, so I think that we will call for a break at this time. Let me compliment all of you on your work and demeanor. Airing the views of two years of your independent evaluations was bound to unleash passions, and it did. That was healthy. And many of you were holding yourselves back from interrupting the person from finishing their say, and I commend you for that. Let us proceed with this same behavior . . . and that includes our discussions between sessions. Remember, think as if we have unlimited time. I am the one monitoring the deadline and I am very certain that we will reach a verdict before October first. Although we currently have a split decision, we will work with that until we get a consensus—either yes or no. Even if we were virtually unanimous today, we would continue on and re-examine until we were completely unanimous. So take your break and let's reconvene at two o'clock."

So the discussions continued. Hour-by hour, day-by day, week-by-week, and month-by month, they plodded forward, oblivious of the looming deadline of October 1, 2034.

September 1, 2034

Mitch awoke at 9 a.m. in his room at NCAR. All of the bedrooms were painted peach orange. The Rev had mentioned to the group that they were painted this color because studies showed that peach orange was a calming color to the psyche.

Works for me. Oh my, I miss Leah. It is still hard for me to grasp that two years can be suddenly compressed into a momentous decision. Ah well, today is the big day . . . and one month ahead of our deadline. May we all work toward the betterment of our beloved and tired planet. Sounds like me! Tired. Mitch rose out of bed.

Their final deliberation would determine the outcome of the launch of the Silver Spacecraft . . . or not. What a long strange trip this has been. In his imagination, Mitch tipped his hat, looking back at the last six months. When The Rev had convened the discussions back in March, they went at it nearly every single day. And the discussions were not always pretty. Viewpoints between Harley's "let's launch now" and Mitch's "let's not launch ever" fluctuated. The Rev rang his time-out chime many times over these months to cool things down. At times, Che-Ling's persuasion drifted toward Mitch. At other times, she shifted toward Harley. She and Harley seemed to hold the key swing votes in the process. At one point, there was a three-to-one poll favoring Mitch's perspective but Hanaford seemed to be wavering. Then Harley offered a new proposal with several conditions that would severely restrict the Silver Spacecraft fleet from ever being permanent. These stipulations included a five-year review period, a requirement that society mitigate CO_2 emissions by 2 percent each year with a binding forty-year deadline and an intensive ongoing monitoring period of hydrologic and ecological effects. Harley's proposal also included an immediate shutdown period if the monitoring demonstrated potentially devastating ecological effects that were 95 percent confident in their probability.

Amid the environmental policy discussions there was also an extensive review of each of their analyses of the validity of the models. In this arena, there was much more agreement. All expressed their confidence in the integrity of the models in their respective fields of specialization. However, Mitch clung to his conviction that despite the impressive work of the scientists who'd created, tested, and verified the models, there was something missing. Most discussions on this subject ended with his saying, "There is a hidden ghost that pervades the very core of these models. That ghost is laughing at us because we are missing the flaw. I need to be convinced that there is no ghost and there is no flaw in the outcome of climate modeling."

Slowly, the momentum shifted toward Harley's proposal. By July, Mitch felt he was up against a wall. Nevertheless, he seemed entrenched in his position, arguing that the stipulations that Harley put on the table were not binding and that once the international community launched the fleet that they would never turn back. All but Mitch were persuaded by Harley's very thorough and convincing technical analysis that the fleet would work. By the end of July, the stalemate worsened and the discussions became heated. The Rev at no time even considered the question of a hung jury. At one meeting, the four scientists unanimously recommended that he bring a "hung jury" determination to Pazderski.

Without angering anyone, The Rev responded. "Patience, my friends, patience. If I inform Pazderski of your recommendation, he will just tell me that it is only *his* call when there is a hung jury. He will tell you to keep deliberating right to September thirtieth. A hung jury will not be considered until the eleventh hour." The Rev took a sip of his water. "Trust me, it will be good. We will find our way. Let's continue to discuss and please, think outside of the box."

In August, two events started to sway Mitch away from his stubborn conviction that the launch was crazy. First was a conversation with Janey after The Rev's "think outside of the box" plea. Janey looked at Mitch. "I know that you are full of convictions and you have made yourself loud and clear. It is your choice on whether to hold firm. However, I encourage you to consider what you learned in your journey through

Door #4. Recall how surprised you were about an outcome for the future of our climate that you never thought possible if society stuck to its Business as Usual scenario of emissions."

The second conversation was an offline chat with Che-Ling during one of the breaks. Che-Ling asked Mitch for some flexibility in his thinking. "Look, Mitch, you know that I am more with you on this. I abhor the fact that we are backed into geo-engineering to get us out of this climate emergency. However, you must think about the prospect of this being a temporary fix. I know your concerns about the 'temporary fix' being a fantasy, but what if we tried to nail down a termination of the Silver Spacecraft fleet after a specified period of time? What if we added a stipulation that after twenty years, the fleet turns off the reflectors? Period. The only process for reactivating the reflectors would be a unanimous resolution of the U.N. to turn them back on."

Mitch thought about this for a few minutes, also thinking about what Janey had said to him the day before. "I suppose that this would take the onus off us. It would imply extreme reservations on our part and our intent that the reflectors be temporary only. If the world overturns our decision, it is the world that will be judged for ignoring the counsel of science." After a pause of silence, Mitch looked up at Che-Ling and said, "Twenty years is too long. Ten is as far as I go." Che-Ling shook her head. "You are one stubborn man, Mitch." He came right back at her. "When it comes to the future of the world, yes I am." They both laughed.

Mitch's train of thought on the last six months was interrupted by Jake, who informed him that the others were gathering in the conference room. Back to the reality of September 1, 2034. In offline discussions all of them had agreed to the principles of the compromise package. He still had major reservations, but he was tired and ready to get this over with. So, this final session should be a formality.

When Che-Ling brought the compromise to the table, they were all ready to vote for a consensus and resolve the arguments. Everyone was exhausted by the deliberation process. Harley made a motion to approve the package that was seconded by Hanaford. Che-Ling added the stipulations that she and Mitch constructed. Surprisingly, there was no opposition to the added amendment to the motion.

The Rev intervened. "We have all come so far. Your discussions have been productive. The debate has swung back and forth for the last six months. However, as one who is removed from the passions of your arguments, I can see you all are coming to a consensus, but partly out of exhaustion. I do not want this to be a consensus from exhaustion. Plus, Mitch has been pinned against the wall on Harley's launch proposal. So, I propose that we take two weeks off from meetings. Look at this as your vacation. I encourage you to use it as such and not work this proposal in your heads or in conversation. Take this as a time to absorb. Then we will have one or more meetings and we will either have a decision or I will start warning Pazderski that we are headed for a hung jury." Somehow, The Rev always had them in hand. He was mostly quiet and imposed a calming influence on each individual and on the group as a whole.

When he did speak, he had their full attention and they always went along with what he had to say. They all agreed to take a break.

So, September 1, 2034 was not the end after all. They still had a month to the deadline. Mitch sensed the frustration of the others, who were ready to vote and adjourn. During the two-week rest period, Mitch took The Rev's instruction to get away from the arguments and used this time for introspection. He avoided talking to the rest of the group and indeed only held casual conversations with Jake and Janey.

September 15, 2034

After the break, he felt like he had gone on a spiritual retreat. It seemed to Mitch that the others felt the same way and were ready to move forward. As usual, The Rev had it spot on—they had all needed the time-out.

So here we are again. I thought we were ready to decide two weeks ago, but The Rev's idea of two weeks of inner sanctuary was brilliant. I kept rolling over the past six months in my head to see if I was missing something. And I discovered what it was and will hold strong to my final conviction. It all hinges on whether the group can accept one more key provision.

Mitch started to come out of his reverie but he fell into a daydream as he imagined dealing with the rest of the jury. *Okay, guys and gals. It is September fifteenth and we are ready to deliberate. The Rev told us that we were not obliged to make a final decision today. He also told us that he was prepared to convey our approval of the launch or inform Pazderski of a hung jury when we take our final vote today. But I am decided, and I hope that you will accept my final position...*

Mitch jumped with a start and fell out of his daydream when Jake came into his room. "What's with you, Dr. Mitch? It looks like you came out of never-never land. We will be convening in half an hour." Mitch nodded. He, had breakfast, and entered the conference room.

Reverend Tyndall asked each of the team to put their perspectives on the question of launching the fleet—or not—on the table. Hanaford, Harley, and Che-Ling, as expected, voiced their support of the compromise proposal. Then everyone nervously awaited Mitch's response.

He took a long pause before speaking. "You know, when The Rev suggested that we go on a break, I fully expected that with the pressure off of me I would return to being entrenched against this proposal. I still think that tinkering with the earth's heat source is a crazy idea. The decision—for society to avoid being trapped in the corner that we are in—should have been made a quarter of a century ago. If this group were meeting on this same proposal twenty-five years ago, there was no way that I would have ever budged. However, I cannot discount the fact that every day as we deliberate people are dying and nations are being sociologically and ecologically upended, due to the climate change impacts in this over-heated world. We did not ask to be put in this position, but here we are, due to the inaction of our leaders in the early part of our century. As I look back at the last six months, I can only admire the process that The Rev has so deftly orchestrated. Hanaford, Harley, and Che-Ling, I hold nothing but admiration for each of you. We argued. We nitpicked. We allowed the passions of our convictions to bleed out onto this table. In the end, it was you three wanting to move forward with the launch and I who was the holdout. But you gave me my space and I appreciate that. As you all know, Che-Ling's compromise of tightening up the 'shutdown' to automatic in twenty years has brought me to a position where I can accept the proposal for our recommendation to launch. Of course, I tightened that timeframe to ten years. But, I want to qualify my affirmation with a significant reservation and I want it noted for the record. First, I have had doubts whether the work we have done will be taken as seriously as they claimed it would be. If we recommended a no-launch, would they have launched anyway? We'll never know the answer to that question, directly." The group was quite moved by all Mitch had to say.

"However, there is a way for us to know whether they took us seriously or whether this was just a rubber stamp committee, and that is at the time when the shutdown of the reflectors is supposed to take place in ten years. If society does shut down the fleet and institute the 80 percent cuts in emissions as we propose . . . or even if the world makes significant progress and asks for a short extension of a few months for the reflectors and then takes them down, I will be pleasantly surprised. If they do not shut down the reflectors, or do so for a limited time period just to have the U.N. rubber-stamp a vote to turn them back on again, then we have been had, and I want this known to the world. So, my vote to launch is conditioned on the automatic issuance of this letter signed by all of us that would be sent to the leaders of each member nation and the major broadcast and internet media if and when this breach occurs."

Mitch handed the draft of the letter to each of them around the table. They took their time to read it and there was a long silence.

Harley then whistled. "This is pretty heavy, Mitch."

He responded. "This whole process has been heavy. You should all think about the consequences if we do not issue it, if and when the time comes. Since we have been pumped up as the 'final arbiters' for this launch, if they renege in ten years, we will have been had and history will look at us as the scapegoats for the negative consequences of having the reflector fleet permanent."

Hanaford responded, "*If* there are negative consequences."

Mitch felt his anger rise. Then he looked at Jake and Janey and bit it down. "Yes, Hanaford, I grant you that there is an 'if' . . . and if there are indeed no negative consequences, this letter will mean nothing other than an asterisk that four scientists evaluating the technology had doubts about it. But if I am right and society keeps controlling the sun ad infinitum and the climate system breaks down into ecological collapse, at least we will be on record, if we survive."

Harley was the first to acknowledge. "I respect your viewpoint and I am prepared to accept your stipulation. But I wonder if the president and other world leaders will accept this stipulation, ignore it, and just order us to keep deliberating or just shut our committee down? What do you think, Rev?"

Reverend Tyndall looked each and every one of them in the eye. "This is your deliberation and I will respect it, even if it is loaded with this letter of protest to the world, in the event of a default on your decision. In any event, what does it matter? You will have done your job and the ball will be in the court of the world leaders. But all of you need to have the unfettered conviction to sign it. If we do not agree on signing the letter, it means that we need to continue to meet to resolve this, and we do not vote today."

Che-Ling offered her thoughts. "I think that Mitch's letter is courageous and should be part of this approval. I am prepared to sign it or vote against the entire proposal."

Harley broke in. "Was this letter schemed up by you and Mitch during the break?"

Mitch answered him calmly but firmly. "I never spoke to Che-Ling or any of you during our retreat away from these meetings. But yes, I did speak to her in August, before our break, and we discussed how this could move forward. It was her idea for a twenty-year deadline to remove the temperature-dampening shield of the spacecraft. I just brought that down to ten years. But never did I discuss this letter as a stipulation."

Harley sighed. "I guess that my initial reaction was to take this personally. I was the first to endorse the launch based upon the engineering, as I truly believe the system will work. However, questioning the impacts of the Silver Spacecraft launch on the earth is out of my league and in your hands. I need to bend to your extreme concerns. The system may indeed work but if the three of you are so concerned on the environmental effects, who am I to stand in the way? The world needs to take our hard work seriously. I am on board. I will sign the letter."

An air of relief filled the room as they all smiled.

"So," said Hanaford, "let's get on with the vote. I move that we recommend the deployment of the Silver Spacecraft fleet, based upon our review of the engineering feasibility and the environmental impact analysis of the project by our team. This recommended approval is conditioned by the following stipulations . . . " Hanaford read all of the conditions of the proposal, pausing slightly before reading the condition that the attached letter be released if the project is not terminated after ten years. It can only be kept running if the U.N. certifies that the system is working and that there are no significant negative environmental impacts during the ten-year trial period and that the 80 percent emissions cuts have been initiated to replace the reflector fleet in forty years with no further extensions."

Reverend Tyndall spoke. "Do I hear a second?" Che-Ling seconded the motion. He then instructed the group. "The seconding of this motion allows and encourages discussion before the vote. Please do so." The group reviewed all of the aspects of their proposed approval. There was some discussion about moving the ten-year deployment period to fifteen years, but Mitch stood his ground.

"I will respect your decision to do this, but that will cause me to vote against this motion," said Mitch.

Everyone was quiet. Harley agreed to keep the deployment period to ten years.

Then Mitch laid a bombshell on the table. "I do not want wriggle room on turning the system off. I object to the U.N. having the ability to reverse the turn-off provision by a 'certification process' as part of the motion that is on the table."

A heated discussion ensued and The Rev rang his chime and asked for a short recess.

When they returned to the table, a final compromise was hammered out. The system is turned off in ten years and only a *unanimous* vote by all nations of the U.N. could reverse the decision and turn the space shield back on. Eventually, a period of quiet descended upon the group. Mitch wondered if they were just exhausted or if they were satisfied that they had come to the right decision. He read The Rev's face. *This is the real thing. He would delay another day or more if he did not think that this was a genuine consensus.* Mitch felt at peace.

Che-Ling spoke. "I call the question on our motion to approve Hanaford's original motion with the additional stipulation that a reversal of our ten-year turn-off period can only be overturned by a unanimous vote by the U.N. allowing an extension not to exceed ten years. Hence, the launch of the Silver Spacecraft fleet for a ten-year period will be temporary, to give the world time to initiate the 2 percent-per-year CO_2 mitigation reduction plan. My motion is also contingent on all of us signing the protest letter, today, and that it be issued if the Silver Spacecraft fleet is not taken down in ten years unless there is a unanimous decision…." Hanaford seconded Che-Ling's motion.

As expected, the vote was unanimous. The science jury recommended deployment. After the vote, the tension in the room broke and there was a feeling of relief. They hugged, high-fived, and enjoyed a period of refreshments and easy conversation. Harley came over to Mitch and shook his hand. "Man, you are one hard-assed juror."

"Ha! I enjoyed sparring with you, Harley. Our opposing perspectives turned out to be healthy for our ultimate decision."

Harley proposed a toast. "To The Rev. We never could have gotten through this period of deliberation without your calm, guiding hand."

All in the room replied, "Here, here!"

Eventually, Jake, Janey, and Mitch broke away from the others. Janey hugged Mitch and Jake slapped him on his back. "Man, you did so well . . . "

Janey finished Jake's thought. "We are so proud of you."

Mitch flushed and blinked back tears. "Well, I guess we will have to wait a few weeks to see how the world receives our recommendation and ten years to see if the world leaders took our concerns seriously."

Jake smiled. "Well, Dr. Mitch, you will actually see that from the other side of the next century."

"Oh, no," said Mitch. "Here we go again." A window appeared in front of the three of them with an opening of glittering silver. They jumped through.

2162

Mitch knew what he was going to see on this reentry to 2162 (same channel different time). Jake, Janey, and he were again suspended in the ethereal cloud observing this dimension. But he did not see Leah-Ann Warner Stuart. Instead, he was looking down on a kitchen where a woman, who looked like Leah-Ann, was cooking a Thanksgiving meal. Mitch put two and two together and recognized her as Leah-Ann's sister. Then Leah-Ann walked into the kitchen.

"Hey, Norah, I see you're working your magic on Thanksgiving."

Norah looked up. "Hey, sis!"

Leah-Ann came over and kissed her sister on the cheek. "Can I be of help?"

Nora smiled. "Yeah, it is time for the bird to go into the oven. Can you put it in with the fixings and set the oven for four hours at 350 degrees?"

Leah-Ann walked over to the pan with the turkey, shaped a square with her hands, and the hologram of a translucent box appeared in front of her. She entered the cook time and temperature on the keyboard and the oven door opened. She basted the turkey and put it in the oven. The hologram door closed and a light hum indicated that the bird was starting to cook.

Nora wiped her hands on her apron and smiled at her sister. "So what's new in Portland?"

"Oh, things are pretty good," responded Leah-Ann. "Tom is working hard toward the end of his doctorate. Sorry he couldn't come today. He is in Syracuse this weekend …

something to do with his doctoral research. And my boy Mitchy? He keeps growing and growing."

Mitch the elder smiled as he heard this and whispered to Janey, "Another branch to the tree of my family."

"No need to whisper, Dr. Mitch. They cannot hear or see us. But congratulations, great-times-seven grandpa!"

Leah-Ann then raised her eyebrows. "You know, Mitchy was again asking about the story of our ancestor from the twenty-first century. I put him off again. I could tell him what I know, but you know more and can tell the story in a more . . . colorful manner."

Nora laughed. "Me, the great storyteller - the wise elder who passes along the wisdom of her great ancestor. Sure! Bro Jack actually asked me the same favor regarding his twins. And my Derrek and Diane always love hearing the story again."

"Wonderful!" said Leah-Ann. "Maybe after dinner?"

Nora smiled again. "I was thinking the same thing."

Leah-Ann helped her with the salad. "By the way, how are Derrek and Diane doing?"

Nora shook her head. "Oh, that rebellious age of adolescence! I see a lot of you in Diane. As headstrong as she is, she is aggressive and ready to take on the world."

Leah-Ann smiled, "Headstrong? Me?"

Norah shook her head. "Oh, let's not get into that again. After all, look how you turned out."

Leah-Ann then ran a hand through her blond hair. "Okay, I guess I am headstrong. Jack would back you up on that."

"No, what I meant was—"

"I know what you meant."

Leah-Ann continued, "So, tell me about Derrek."

Nora replied, "He was just accepted into Cornell. He still wants to follow in your footsteps into the environmental science field."

Leah-Ann smiled. "Oh, I'm flattered. Cornell is a great college for Derrek to get his start in his education. I look forward to talking to him about it."

Nora looked relieved. "I would love for you to talk to him. You know how I feel about this."

"Now, Nora, I am not going to try to talk him out of his first idea for a future career. I will just give him the ear of an interested aunt."

Nora started the vegetables. "Oh, all right."

The doorbell rang. "That must be bro Jack and his family. I'll go greet them."

Nora responded, "You know how much he hates being addressed as 'Bro,' don't you?"

Leah-Ann laughed. "I do," she replied and went to the vestibule to greet her brother. "Jack! How are you, Bro?" She gave him a hug.

"I'm doing okay. Still unemployed but I have a few hot prospects including an interview with the NCAR Geo-Engineering Center next week."

Jack's twins Betty and Eddy pecked their Aunt Leah-Ann on the cheek and ran past them to join their cousins in the living room.

"That is wonderful, Jack. There is so much going on at NCAR!" said Leah-Ann.

From above, Mitch tried to scratch his beard, forgetting again that he had no beard. "About time. I thought I would never hear anything about the climate of this dimension. Wait a minute . . . did he say the 'Geo-Engineering Center'? You mean that damned Silver Spacecraft fleet is still up there?"

"Wait. Wait," said Jake. "You will find out in due course."

Janey followed up. "Dr. Mitch, didn't you get the clue from Nora and Leah-Ann? Remember, the storytelling period after dinner? That's when you will learn what happened."

Mitch nodded. "Oh yeah . . . but can't they sit down and eat already?"

Jake and Janey laughed.

Grandpa and Grandma Warner came in and joined the party. Leah-Ann told them about Nora's storytelling time planned after the meal.

"Splendid!" exclaimed Grandpa Warner. When they were all seated at the dinner table, Leah-Ann announced that Aunt Nora was going to tell the full story of their ancestor, Dr. Mitch, and his battles against the geo-engineered space shield.

"Oh, boy!" cried Mitchy. "He has the same name as me. Was he a great man?"

Leah-Ann turned to her son. "You will just have to wait. Aunt Nora tells a good story."

After the meal, the family gathered in the living room. The younger generation was restless. Leah-Ann's son, Mitchy, kept asking when Aunt Nora would tell her story.

"Mitchy, sit down. We have to wait for her to finish what she is doing in the kitchen," said Leah-Ann. At that moment, Nora entered the room. Amid pleas from the five cousins to start her story, she served drinks and cookies to the gathered family.

Grandma Warner hit her cane on the floor. "Come on, Nora, tell the damn story!"

Nora sat down on her favorite chair and sighed. "Okay, okay." She took a deep breath and began.

Mitch looked down on the gathered family of his descendants and yelled. "Yes! Come on, Nora. Tell the story!"

Jake and Janey laughed. "I think she may have heard you, Dr. Mitch. She is getting ready to begin," Janey said in jest. Mitch laughed, then he became intent on the beginning of the story.

"Well, first of all, we do not know much about our family tree before your great ancestor Dr. Mitch, other than the names of his parents, Glenn and Marianna. That is why one of you needs to continue the family record that we have started after your parents, aunts, and uncles are gone. Our family story is weaved largely around your great ancestor, Dr. Mitch."

"Was Great Grandpa Mitch a doctor?" asked Mitchy.

Leah-Ann looked at him and responded to all of the cousins of the next generation. "Now kids. Let's try to hold your questions until Aunt Nora finishes. Her story is beautiful and you will like it if she just keeps telling it to you. And Mitchy, your great ancestor Dr. Mitch was a doctor of science, not of medicine."

Mitch, watching the story from a cloud, yelled, "Atta-way, Leah-Ann, let's get this show on the road!"

Janey poked Mitch. "Shhhh! She's starting again."

"Our story is as much about our climate as it is about our ancestors. And the story of our ancestors goes beyond what I am going to tell you. Let's consider this 'story one' and I will fill you in on other ancestral adventures of our family another time. But as you hear me, you will see how great great—aw let's just call him Great Ancestor Mitch—influenced history. This is the shining story of our family history. In the early period of his career, Great Ancestor Mitch was a professor of climate science at the Miami Center for Environmental Research. We are told that he was a great storyteller for his classes and entertained them in a manner that his students loved. They wanted to take as many classes with him as possible, which was why the classes were always over-enrolled. One of the stories handed down about Great Ancestor Mitch is that he once took his class on a boat in the Atlantic to snorkel and observe the dying coral reefs. During their snorkel observations, one of the students drifted out too far and when Dr. Mitch swam out to bring him back to join the group, he saw the student fleeing from a shark. The student had cut his wrist on one of the reefs and the blood must have attracted the shark and it was heading in for the kill. Dr. Mitch swam in front of the fleeing student and as the shark charged at him, he drew his harpoon gun and zapped the shark, hitting the poor hungry creature square in the head and ending the danger."

Mitch laughed in the cloud above the family. "I never did that. I wonder who spun that yarn?"

Janey replied, "Maybe you did do it, Dr. Mitch. Remember that this is a different dimension from your 2020 world."

Nora continued. "Another story involves his practical jokes. He was so colorful—both literally and figuratively. He came to class every day with a different type of Hawaiian floral shirt. As for his practical jokes, he once gave each of his fellow faculty members a Hawaiian shirt that fit them. He asked the entire faculty (including the dean of his school) to wear them the next day for the faculty meeting—and told them to wear them with no undergarments below the shirt. He had ordered them from a company that sold practical joke materials. These shirts were designed to bleed through to the skin after a few hours of wearing and they all did. Apparently the dean wore it half the day, and then decided to change back into his suit after making his showing in the morning at the Environmental

Science faculty meeting. When he took off his shirt, he was shocked that there was a floral imprint all over his upper body, like a tattoo. The dean immediately called the department secretary and told the faculty to gather in the faculty lounge at three o'clock. He entered the lounge at three, took his shirt off in front of the faculty, and said, 'Mitch, this time you've gone too far.'"

"The other professors then checked their skin and all agreed. 'Mitch, this time you have gone too far!'"

Everyone in Nora's living room laughed with glee.

From above, Mitch was also laughing. "Now *that* story is completely true." Jake and Janey just shook their heads.

Nora turned serious. "Dr. Mitch was a talented climate ecologist. He did a lot of research and published articles on the coming collapse of the climate system and how it would generate a collapse of ecosystems throughout the world. In the period leading up to the 2020s, he was very involved in the debate about whether society needed to go on a crash course to reverse our CO_2 pollution of the atmosphere by heat-trapping gases. His personality changed when he involved himself in advocating for an 80 percent cut of CO_2 emissions. His levity disappeared and he became intense and argumentative. He spread his message through his research, publishing many reports of his modeling of how ecosystems would start to collapse with every degree of temperature rise that occurred. Society did nothing to stop its pollution. Then in the 2100s, long after his passing, Dr. Mitch's predictions came true. Temperatures started rising rapidly and almost every one of his climate models—each focusing on a different ecosystem—came true. He was wrong on the coral reefs. He predicted that they would be gone by 2100. They disappeared by 2025. As the world entered this climate emergency, they turned to the Silver Spacecraft fleet to reflect sun and quickly lower our temperatures to stop the disasters. Does everyone know of the Silver Spacecrafts?"

Most of the younger generation said, "Yes."

Mitchy said he'd heard of it, but did not know how it worked. Derrek said he would explain it to him after the story.

Nora continued. "In ten years, the UN designed and engineered the fleet of spaceships that would reduce the sun's incoming heat within five years to lower global average temperatures. This would reverse the warming and end the emergency. However, there was worldwide protest by many people who feared that we would mess up our planet's climate by artificially reducing sunlight. Dr. Mitch was in the forefront of that protest movement. So, the world decided to hold off the launch and convened a council of UN supported scientists and engineers to make a final evaluation of whether the fleet would, indeed, work. They were referred to as the 'science jury' and were tasked with making a recommendation on whether or not to launch the Silver Spacecrafts. A second equally important question would also be addressed: could the fleet reflectors operate without unforeseen environmental side effects? Ancestor Mitch was appointed to this jury. Allegedly, Mitch was the lone voice in recommending a stop to the fleet. He was outnumbered by others who said that the launch should go forward, but he single-handedly crafted a compromise forcing the world to turn the mirrors off after ten years to give us time to reduce emissions and bring in a natural process for a cooler climate."

Mitch smiled from above. "Well, that's a bit of an exaggeration, but I'll take the glory."

Janey patted him on the shoulder. "Brace yourself for what comes next, Dr. Mitch."

"So, after the recommendation of the geo-engineering/environmental jury, the Silver Spacecraft fleet was launched. And, as predicted, it worked like a charm. Within two years, the average annual temperatures started on a downward trend. By year five, the rate of temperature decrease was so apparent that the emissions of nanoparticles on the spacecraft fleet were adjusted to let a little more incident sunlight come through to the earth. It was like an adjustable shade—they used to call them 'venetian blinds'—to let more or less sunshine into a room. They did have to continue to adjust the space shield emissions of nanoparticles over the years to come. The temperature by year ten had settled into equilibrium." Derrek interrupted, "What about the environmental side effects and what happened to the shutdown at year ten?"

"I was just getting to that," said Nora. "The world watched nervously. During the first five years, there were conjectures in different parts of the world about less rainfall, and colder winters, but nothing significant was noted on the climate scale. Climatologists explained that the reports from different corners of the world were based on regional weather patterns, not the big picture of climate. In fact, climatologists were warning the public that it would take at least another twenty years before a good judgment could be made about whether these effects were due to climate or if this was just a temporary weather trend. Then in 2041 there was a very weak monsoon season, almost nothing to speak of. Thereafter, the monsoons were gone. This caused great concern, because the breakdown of the Asian monsoon was a red flag predicted by the earlier climate model projections. Again, the climate science community warned that it was way too soon to make a judgment."

Mitchy complained to his aunt, "I want to hear about Grandpa Science Doctor Mitch." Everyone laughed.

"Okay. This is where our story gets interesting," said Nora. "When 2044 rolled around, the temperatures were consistently lower and approaching average temperatures of the pre-industrial revolution before humans started emitting CO_2 from vehicles and industries. The American President's Science Advisor told everyone to hold on as we still needed twenty more years of data to say anything statistical about the effects of the space shields in cooling the climate. However, he also mentioned that the annual average temperatures of nine of the last ten years were much lower than the previous 170 years and that this was a good sign. So here is where Ancestor Mitch, comes into the picture. The U.S. and the world knew of their commitment to turn off the space shield permanently and continue the hard job of reducing emissions. In fact, emission cuts were only one percent per year in that decade instead of the required two percent per year. But in the end they reneged. They were prepared for the end of the ten-year period and they appointed a different geo-engineering/environmental review team of scientists to make another evaluation. This, of course, angered Ancestor Mitch. He smelled a rat."

Dr. Mitch, from above the room turned red, his cheek twitched, and he clapped his hands in anger. "Damn it! I know where this is going and I can already say, 'I told you so!'"

Jake and Janey both said, "Hold on!"

Jake followed up. "There's much more to the story. Wait."

Nora continued. "When the space shields went down, there was a worldwide outcry. Most people felt that the Silver Spacecraft fleet was doing a great job and they should go back up. Many nations of the U.N. also started objecting to taking the spacecraft out of commission. Just as in 2030, China, Japan, and Brazil had warned that they were ready to buy out of the world treaty and launch a fleet of their own. The pressure was intense. The new science policy review team finished their evaluation in six months and recommended 'go.' The U.N. approved the recommendation by a less-than-unanimous vote. The temperature-controlling shield of nanoparticles was reactivated. Our ancestor, Dr. Mitch, immediately lashed out. He contacted the other scientists of his 2032 environmental review council. He learned that Dr. Hanaford had disappeared on an African expedition. Dr. Harley did not return his calls. So Dr. Mitch and Che-Ling released the signed letter. It was issued to every member nation, the internet, and to all news media. It made a big splash on the internet. Dr. Mitch and Dr. Che-Ling even hacked into the hologram universal broadcast system and presented their own case that the world community had not lived up to the conditions of their geo-engineering/environmental jury. Someone (or some group) had spray-painted a very large graffiti on the face of the U.N. that said "Liars!" There was no evidence that Ancestor Mitch was responsible, but he was pretty steamed and the media played it up as his work of graffiti art. The world's failure caused a news splash over a few weeks, but this eventually faded."

"Did I do that?" said Dr. Mitch from above. "Wait a minute! I am getting screwed up again. How could that be me in the 2040s when I would be close to eighty?"

"Whoa, whoa, Dr. Mitch," said Jake. "We'll explain in due time."

Janey added, "When this story finishes we will explain all of that. Remember how you look in the mirror. For now, just Let it Be."

Down below, Jack's son, Eddy, interrupted the storyteller. "So did our Ancestor Mitch just fade away, too?"

"That would be a sad end to his role in our history," added Betty.

Nora smiled. "Yes, it would have been sad, but there is a twist in this part of our family history. By this time, Dr. Mitch was in his early eighties. He was very bitter and retired from his profession. Then something interesting happened. His grandson Martin . . . your great-times-four grandfather . . . started following in Dr. Mitch's footsteps. He trained as an environmental ethicist and environmental policy analyst. He did his doctoral studies in climate modeling and how it influenced the decision of the 2034 geo-engineering/environmental review council. His doctorate drew a response from those in the world community who still felt we'd made the wrong decision in *not* shutting down the Silver Spacecraft fleet in 2044. In the meantime, Dr. Mitch became so proud of his grandson that he started mentoring him. Then, the climate started behaving strangely. The temperature plummeted more seriously in the 2050s, and studies showed that the lower incidence of sunlight was amplified by the colder ocean currents. Winters in the mid-latitudes extended to eight months out of the year. Among other things, this caused the world's breadbaskets, which had already shifted to the upper latitudes, to break down in certain regions. Agriculture shifted south as temperatures plummeted more than predicted. Then the Asian monsoons disappeared and there were more years without the rainfalls that supported mid-Asia's farming. At the end of the decade, climate scientists announced that there was enough statistical data to attribute these weather patterns to a climate phenomenon caused by the Silver Spacecrafts."

"So, our Great Ancestor Mitch was right!" interjected Derrek.

Nora looked at him. "Well, let's see where our story takes us from here. The international community, after an engineering evaluation, recreated the design of the nanoparticle release to make micro-adjustments to emit just the correct amounts of the particles as needed that would automatically change the amount of sunlight blocking as needed."

"In the meantime, Dr. Mitch teamed up with his grandson Martin and Dr. Che-Ling. He let his grandson Martin take the lead in organizing an interdisciplinary movement to phase in controls of CO_2 emissions that would ramp up to the 80 percent cuts that were always advocated to be the permanent fix. They adopted a slogan: 'The Silver Spacecraft fleet was always intended to be a temporary Band-Aid.' Old-timers tried to dub their initiative 'Let the Sun Shine,' after a song from a 1970s musical, but it never took. Instead, it was named simply The Sunshine Movement. Grand ancestors Mitch and Martin struck a chord worldwide. Scientists from each nation organized their own 'Sunshine' groups of citizen activists. Dr. Mitch and Martin created the oversight organization and they became known as the 'Sunshine Boys.'"

"I don't remember doing that," said Dr. Mitch from above. "Wait, how could I? Unless you never took me back to my world of 2020!"

Janey rubbed Mitch's neck. "No, Dr. Mitch, you will be returning to your world soon. Just bear with us and we will explain all of this to you after the story."

Mitch sighed. "Okay."

Below, the family was entranced with Nora's story. "It sounds like the world was really falling apart back then. Our history books gloss over that," said Derrek.

Nora responded. "That's a story for another day. In the meantime, let's move on. By the early 2060s, the adjustments had been made so that the reflector shields could self-monitor according to a program that would gradually bring the temperatures back up, then constantly sense the radiation inputs that would keep the sunlight fluctuating in a pattern that a computer program dictated, in order to reach a more normal climate. In other words, instead of making adjustments when they were needed, the shields would be adjusting all of the time. Although this fix was widely supported by the engineering community, it was widely opposed by mainstream ecologists. In the meantime, most citizens of the world were frightened enough to jump on to the 'Sunshine Boys' bandwagon. That is when the U.N. invited Dr. Mitch, Dr. Che-Lin, and Martin to address the world body to present a proposal. Their proposal showed a way that we could phase in 80 percent emission cuts, with the goal of eliminating the Silver Spacecraft shields by 2100. As the emission cuts were instituted, the reflectors would be phased down by 20 percent per decade. By the end of the century, the world would abandon the shields and let nature take its course. The world was so turbulent from this new climate emergency that it eclipsed all other global issues. There was a wave of support for the 20 percent per decade emission reduction proposal framed by the Sunshine Boys. Amazingly, the U.N. adopted the proposal and all of the nations signed on with their own regulatory systems to do so. The business community was in full support of the Sunshine Movement as they realized that the world market needed stability and returning to a natural climate would achieve that. Industries pursued crash development of technologies for more mass transit systems, bike lanes, electric vehicles, and the phase-out of fossil fuels."

Derrek interrupted. "So, our Grand Ancestor Mitch was right all of the time!"

Betty added, "This is big stuff. I do not remember reading this much detail in our history classes."

Nora responded. "I'll get to the recording of these events in history, but let me address Derrek's comment. Dr. Mitch, in his later years, was interviewed by the World News Network along with Martin and Dr. Che-Ling. Grand Ancestor Mitch was asked if he felt vindicated and whether the geo-engineering solution should never have been deployed. Ancestor Mitch responded to that question in an interesting way. 'In retrospect, I stand by the recommendation of our science review jury of 2032–2034. We needed a temporary fix to the climate emergency of that time. But we should have immediately tackled the phase-out of carbon emissions that we are undertaking now and we should have turned off the shields in 2044, as was recommended. Yet, for all of my stubborn objections to this technology, in the twilight of my life I see its value. Back in the 2030s, we had a chance to get it right. We had a climate emergency upon us and we needed to do something to give us some breathing room. Had we deployed the spacecraft with the absolute intent that this would be a temporary solution, the temperature shield of nanoparticles would have served a noble purpose. Now we have another chance, if the world seriously bites the bullet and allows our Sunshine Proposal to phase in 100 percent renewable energy and phase out the Silver Spacecraft fleet.' "

"The twilight of my life?" said Mitch from above. "She is not going to tell them of my death or the year of my death, is she?"

Jake was impatient with Mitch. "No, Dr. Mitch. Keep your pants on. She's almost done. And we will tell you why it wouldn't matter to you even if she did describe your death in this dimension."

Nora started drawing the story to a close. "Unfortunately, our Great Ancestor Mitch never saw the results of the Sunshine Proposal. After his death, Martin and Che-Ling were retained by the U.N. to shepherd in the emission cuts process and the phase out of the space shields. Amazingly, they did! By 2110, the world achieved 60 percent emissions cuts—75 percent of the target for 2100—and the shields were adjusted down to a minimum. We continue to cut our CO_2 emissions and are approaching the 80 percent that was our target for 2100. Better late than never, I guess. Do you think that your history books recorded these dramatic events from the past 130 years?"

"I think I remember that vaguely," said Eddy.

"Oh, they just gloss over that," said Derrek.

Mitchy added, "I remember asking my teacher about it last year, because Mom told me that our Great Ancestor Mitch had something to do with it. But my teacher shook her head and turned to another topic."

"It's a shame," said Nora. "I finish our story with another bit of family history. As your great-times-four Uncle Martin was getting on in years, he was assisted by two of his nephews, who he dispatched to Central Africa and Japan to assist communities in sustainability practices. The battle for a clean, natural planet went on. Society really took this seriously and now we are living in a new world. Our climate was permanently moved off of its course from the track we were on before the last three hundred years of the nineteenth-century industrial revolution and the ill-conceived decision to keep the Silver Spacecraft shields on in 2044. We still have more extreme events, but your Aunt Leah-Ann can explain to you that this is a 'planetary growing pain' as our planet evolves into a new and hopefully more stable era."

"So, you have questions, I know, but the one that is bursting out of you is why recorded history is vague about these big events that shaped our world over the past century. So, let me tell you what I think . . ."

As Nora was wrapping up her storytelling, the scene started to fade from where Mitch, Jake, and Janey were sitting. Mitch was crying as he looked one last time at his descendants. Jake and Janey also had tears in their eyes. As the air around them became foggy and wavier, the bubble turned a lighter shade of silver.

Janey said, "We now move on to the next door of this journey. We will have a few minutes to talk."

Jake followed up, "Dr. Mitch, we owe you an explanation. We need to bring your head together about the Dr. Mitch that you just observed—his age, his experiences and the end of his life—and reconcile him with the Dr. Mitch who you are."

Mitch jumped in. "It is starting to become clear to me. I am Mitch from the year 2020. The Mitch we heard about in this story was a different Mitch from the mid-2000s. I cannot connect the two Mitches, so you are going to advise me to 'Let it Be.' It's hard, though, because I am seeing the other Mitches in so many different possibilities and Nora's description of the end of my . . . or his . . . life has thrown me for a loop. How am I to return to my world of 2020 without knowing which of these futures my life will take? I feel like I have lost control over the different scenarios that will affect my life . . . and I don't know what destiny my 2020 world will take me to. I guess I will have to . . . let it—"

"No, Dr. Mitch!" Janey broke in. "Don't Let it Be! You were on track for a while when you were looking at the worlds of the different doors as different alternatives of where we might go. Different 'dimensions,' as you have been calling them. You have now slipped back into thinking that the Dr. Mitch that you just heard about is the same Dr. Mitch that you were in 2020. You saw the incongruities and I don't blame you for being confused."

Jake continued. "Dr. Mitch, have you ever heard of the 'multiverse' or 'multi-universe' theory?"

Mitch thought about this. "You know, there was a physicist in my time named Stephen Hawking. He studied the 'string theory' that spins off to an explanation of the possibility that we are living in a universe that may not be the only universe. It is based upon the work he and other collaborators developed in the field of quantum cosmology. Oh, I am speaking way too technically here. Forgive me; I am actually talking to myself as I'm figuring this out."

Mitch took a breath, then continued. "The theory states that, there are different possible universes. The universe that we are experiencing is just one of many. The various dimensions that are out there are sometimes called parallel universes. So, each of the doors that we have been journeying through is a parallel universe . . . portraying what happens in its own dimension, but not necessarily what happens in my world. Each of the people that I see and interact with is in a different dimension of possibility of what might happen. Am I on track?"

"Bravo!" said Janey.

"You are *mostly* on track," Jake added, "but your idea of a different possible outcome is not quite right. In this dimension, you are in a future period of a different universe. Same channel, different time. The world you came into from your dimension of 2020 is, as we speak, proceeding down a different path, separate from the one that you just heard about from Nora."

Mitch asked, "But what about me in these different worlds? Am I the Dr. Mitch from my world of 2020 or am I the Dr. Mitch of the world that I am in right now?"

Janey responded. "When you are observing, as we did from the cloud above your descendants in 2162, you are in your 'world of 2020.' When you are interacting, as you were in the geo-engineering/environmental review jury that recommended launching the fleet, you are in *their* world."

Mitch scratched his non-existent gray beard. "I am starting to get it. But what about you two . . . and Casey? What universe are you from?"

Jake replied, "Now you are getting ahead of yourself. It is time to Let it Be. All will become clear by the end of your next journey of this evening. One more door."

As they time travelled to the next era of Door # 5, Mitch suddenly heard a song from the 1970s that resembled his experience through Door #5. He saw a man with long stringy hair playing a guitar in the silver haze. Yes, it was Neil Young singing something about a gold rush; something about loading silver spaceships to a new home in the sun.

Endnotes: Chapter 5:

1. The concept of geo-engineering is not science fiction. It has already moved from fiction to a place in respectable engineering and policy journals. The technology is rather simple and is explained in the exhibit displays described earlier in this chapter. As Jay Michaelson writes in his paper "Geo-Engineering: A Climate Change Manhattan Project".[38] "Once the research into geo-engineering has yielded feasible results (if any), the second phase of a geo-engineering Manhattan Project would involve the development and deployment of the most desirable geo-engineering proposal that emerged from the first phase. It is difficult to know now exactly what the Manhattan Project would finally look like."
2. One estimate of the cost of deploying geo-engineering fixes is $100 billion and annual operating costs of $20 billion per year.[39] Geo-engineering at this time in our society is met with stiff opposition from environmental groups and is considered by some to be folly. However, the premise of this chapter is that climate change effects cascaded rapidly in the 2020s and society was backed against a wall, and fully immersed in a global emergency. In such circumstances, geo-engineering could be a "quick fix" option. The engineering concepts are realistic and could be as simple as dumping iron filings in the Southern Ocean, as described in the "museum" section of this chapter. But, the more likely deployment would be emitting aerosols from spaceships into the stratosphere. Before proceeding to deployment, a massive study of the environmental effects needs to be considered.
3. Our journey through Door #5 takes us through a simplified process involving a jury of scientists and engineers who review all of the existing studies and recommend a policy decision. But should geo-engineering ever be considered, given its unknown long-term environmental effects? To those firmly entrenched against the idea: Can you imagine a world emergency where geo-engineering could be deployed as a temporary measure to keep the world alive while emission-reduction strategies are implemented? To those who advocate deployment of geo-engineering: Would geo-engineering cause radical long-term and unpredictable changes in the world's climate? This is up to the reader to consider, in retrospect of the scenario described in Door #5.
4. According to Wikipedia, the multiverse (or meta-universe) is the hypothetical set of infinite or finite possible universes (including the universe we consistently experience) that together comprise everything that exists: the entirety of space,

---

38 http://www.metatronics.net/lit/geo2.html#two, Introduction, second paragraph
39 See http://archives.ppgbuffalo.org/wp-content/uploads/2010/06/How-To-Cool-the-Planet.pdf

time, matter, and energy as well as the physical laws and constants that describe them. The various dimensions within the multiverse are sometimes called "parallel universes" or "alternate universes." There is a reference in this same Wikipedia definition that links Stephen Hawking's work to multiple universes that may exist in reality.[40]

---

40 https://en.wikipedia.org/wiki/Multiverse

CHAPTER 6

THE CRIMSON DOOR (#6): THE VOLCANO

What happens to the earth's climate once the ash and aerosols of a super volcano eruption disappears?[41]

***

2024

*P*OP! The bubble that they were in ceased to exist. Jake and Janey were back in the vestibule with Mitch. "For tonight, only one more door to go." said Jake. Janey coyly looked into his eyes and winked. Mitch thinks to himself: *I love all this attention but, wow, I don't know what to do with the flirtations. She sure pushes my buttons. Having this more youthful appearance certainly has its upside and its very confusing downside.* Mitch shook his head. He opened the crimson Door (#6) and entered a beautiful, natural setting. They were in a topographic bowl, standing in a relatively flat area that was bound in on all sides by a rising elevation. All around them there were geysers and colorful bubbling mud pools. In the distance they saw a gush of water steaming out of the ground.

"Do you know where you are, Dr. Mitch?" asked Jake.

Mitch took a deep breath and smelled all the sulphur and steaming minerals. "Well, I would have guessed that we were on Mars if I had never been to Yellowstone National Park before!"

Mitch looked around this wonderful place and recalled some of his fondest memories. *I have loved this place ever since I did a thesis on organics and hot pools.* He broke his reverie and spoke to Jake.

"So, okay, humor me, amorphous JJ twins, what are we looking for in this alternative universe and what am I supposed to do?"

---

41 Volcanoes emit large quantities of ash and aerosols that cool the climate. They also emit CO_2 that warms the Earth's atmosphere. For more detail, see chapter endnote 1.

Janey replies. "Let it Be. Close your eyes, take a deep breath and enter into this reality with new eyes." Mitch closed his eyes for a moment and reflected on what other world might be coming his way.

Mitch opened his eyes and looked around. He jumped into his new reality and his active-participant role. He ordered Jake and Janey to start setting up camp while he took one last check of the seismic monitors.

Jake whispered to Janey, "I think that he has re-oriented himself fully to 2024."

Janey replied, "Let me check for sure." She called out to Mitch as he was heading toward the monitors. "Wait, Dr. Mitch." He turned around. "I need to ask you something."

He sauntered over to where Jake and Janey were standing. "Okay, but make it quick. I have a bad feeling about the last two readings of the seismometers."

Janey asked, "First of all, what year are we in?"

Mitch stared at her for a moment. "I know that in this dimension of Door #6, the year is 2024. As you wished, I have just jumped into the Dr. Mitch active role of this reality of the universe."

Janey looked at Mitch, smiling. "And do you know what we are doing here?"

"Oh, I see. This is a quiz. Okay, I'll go along. We are on a critical scientific mission. In recent weeks, seismic monitoring in the area detected unusual spikes of activity, which have increased in frequency in and around Yellowstone." Dr. Mitch could not help but notice Janey's slender muscular legs. He turned away. *Leah. What a wonderful camping trip Leah and I had when we went on our honeymoon. She wore shorts like that, too.* He smiles as he took in a deep breath. But back to reality: "We are aware of the fact that a big eruption of this super-volcano is overdue, so when the seismic activity began to jar the earth, the U.S. Geological Survey recruited me, with you and Jake as my assistants, to study the area to make sure that the 'big one' wasn't about to erupt. For some reason, I have morphed into a geologist in this version of the universe. We are using ground-penetrating radar in conjunction with conventional seismic monitoring devices to test the intensity and direction of these signals. Today we are working at the edge of the caldera. Am I up to speed, Janey?"

"You certainly are!" she said. "You have come far in your understanding and now can just assume your role. So, let's move along."

After Mitch set up the radar and Jake and Janey finished with setting up the camp, they went out to the location where the monitors were staged. He knew the caldera[42] was so huge—encompassing the entire perimeter of the park—that it was impossible to see its shape unless you were in an airplane. He was in awe of the beauty of the landscape as the setting sun illuminated a crimson cast on the otherworldly features: the hot springs, the

---

42 A caldera is a large cauldron-like volcanic depression, a type of volcanic crater (from one to dozens of kilometers in diameter), formed by the collapse of an emptied magma chamber. The depression often originates in very big explosive eruptions. See https://en.wikipedia.org/wiki/Caldera

boiling mud, the geysers, and the moon set a surreal effect to the landscape. He looked forward to having dinner by the campfire with his two quirky friends and turning in for the night.

My mind seems so much more, let's see . . . what is the word . . . vibrant, since this whole time-travel ordeal began. Well, ordeal is not the right word. It is more of an awakening. Yes, that's it! I can feel the subtleties of the world beyond the observable circumstances. Maybe that is why I seem to accept more and argue less. Life is really a dance. I miss Leah. Whenever I see Janey, I think of my darling Leah when she was young, vibrant, and quirky! I hope I have not caused her to lose her vibrancy with my years of blind ambition in my profession. I am definitely taking her to our old camping spots. Yes, we are going to travel across the country and see the sights again.

His dreamy thoughts suddenly dissipated as the earth began to shake. Jake and Janey turned pale as they ran toward where Mitch was at the seismometer and ground-penetrating radar site. Mitch told them that this was probably a routine tremor. He then checked the readings of the probes. With some alarm, he noted that the strong radial patterns of seismic disturbances had an epicenter near the cone of the former crater that had erupted 600,000 years ago.

He called out to Jake and Janey. "We've got something strange on these readings. We better be cautious and get the hell out of here!" Rather quickly, they packed up their equipment and began walking out of the caldera on a trail that led to the gravel lot where they were parked. Each one of them stumbled side to side as if in a drunken stupor as smaller tremors continued. The leaves danced on the larger trees and smaller trees were flowing side to side. All three stopped dead in their tracks as they heard a tremendous shrieking and cackling sound as, overhead, thousands of birds took flight.

"Look!" Janey pointed to the sky as all three froze in awe at the spectacle. "This is not good. Birds have a sixth sense about something big about to occur." Mitch urged them to hurry down the trail, running and stumbling for their lives with Jake in the lead.

"There's the truck!" exclaimed Jake.

As Mitch approached the truck, he calmly reassured Jake and Janey. "We are not out of here yet, but at least we have wheels." As they crossed the parking lot approaching Mitch's 4x4, another tremor, more violent than the first, shook the earth.

As they were getting in the truck, Janey stopped. "Wait, listen!"

Dr. Mitch was impatient. "What is it?" Then he and Jake noticed it as well. It was not the sound, but the lack of it that was striking. They were at the 4x4. "Get in—quick!" Dr. Mitch was now the one feeling the scare of an impending event.

The earth began to shake, more violently than ever. A noise that sounded like a sonic boom of a plane breaking the sound barrier rocked the area. Mitch, Jake, and Janey stood transfixed; a plume of crimson smoke was visible a few miles away.

Jake and Janey turned pale. Janey muttered, "That's no geyser, Dr. Mitch." The earth rumbled again.

"It surely is not!" said Jake.

"Let's go!" said Janey. They drove out of the parking lot and then noticed elk, bison, and a grizzly bear moving northwest out of the caldera.

"I guess we need to take our cue from those guys," Jake quipped.

Janey pointed. "Strike forth, Dr. Mitch, north by northwest." Dr. Mitch laughed and admired the sudden calm of his two mates. Strange! A few moments ago, he'd sensed their panic.

Mitch drove as fast as his truck could take them down the road. Bouncing and weaving on the red dirt road, they exited the caldera and kept driving along the road leading away from the park. Fortunately, at this late hour, there were few cars on the road, but all were racing to exit the park. The adrenaline rush added to the tension in the car and Mitch decided to calm things down. "Okay, mates, let's put this in perspective. This may or may not be the big blast that has been overdue from this caldera." Mitch looked in his rearview mirror. "Oh my God, look at that plume. I never really connected to the real life circumstance of experiencing such a natural disaster. So, let's just ride out this experience as an adventure."

Jake was nervously fumbling with his headband. Janey looked ashen. He tried to calm them down with more facts. "As you know, we are on a geologic hot spot here. However, this eruption may not be as violent as the one that blew 600,000 years ago.[43] Nevertheless, we have an eruption. It may take several hours or even days to crescendo to the maximum blast."

After a moment of reflection, Janey asked, "How far do we have to go until we are in a safe area?" This time Mitch did not sugarcoat his response. "Based on evidence from its last eruption and today's northwest wind, we need to head to Washington State. To be safe from the cloud of ash that comes out of a Yellowstone eruption, at the very least, we need to be out of the state of Idaho. Jake, get the GPS going and plug us in toward any address in Seattle or Spokane. Janey, you man the map and navigate."

Three hours later, they were out of the park and halfway through Idaho toward Washington. It remained ghostly quiet, but that silence came to an abrupt stop. A deep, slow, rumbling sonic boom bombarded their ears, as they looked back toward the direction of Yellowstone, they saw a black and red mushroom-shaped plume. The earth was shaking hard.

"My God!" exclaimed Janey. "Drive faster! The ash and the fires have lit up the sky to the east!" The night sky was filled with color that was magnificent and equally destructive. Janey looked anxious. "Are we safe, Dr. Mitch, are we safe?"

The professor gave an unsettling answer. "I don't know. I just don't know." Janey reached over to the dashboard and turned on the radio. Only a piercing static was audible. They were all silent as they sped on.

---

43 See http://www.yellowstonepark.com/natural-wonders/volcanos/.

Six hours later, as they passed out of Idaho and into Washington, a slow-moving rumble on the road suddenly began shaking them violently. Dr. Mitch let off the gas and began braking as the tires began skidding and the truck began bouncing up and down as well as weaving side to side. "Oh boy . . . hold on! I'm going to lose control soon." The next few moments seemed like an eternity as the truck weaved violently until the last moments before the last wave of quakes lifted the truck up in the air. They rolled three times and landed on the driver's side, sliding another 100 feet into a field. The windshield popped out as a barrage of summer wheat slammed against the side of the car and into the truck. When it stopped, they all sat still for a moment coughing and moaning as the earth quieted down somewhat.

Jake asked "Is everyone okay?"

Janey responded: "Yes, I think I am, but look! Mitch needs help!" Janey quickly responded. The twins unbuckled their belts and guided themselves to help Mitch, who was moaning in pain. The ground was still shaking slightly. Their world seemed to creep along as in the slow motion of an accident as they made their way toward Mitch. They pulled him out of the side widow that was facing upward.

As they got Mitch out of the car, Jake suddenly needed some tangible answers "Are we going to be okay, Dr. Mitch? Are we far enough away?"

Mitch was abrupt. "Damn it, I don't know. We need to take this one step at a time. I think my wrist is broken—are you both okay?" Jake responded for the both of them that they were uninjured.

Janey looked pensively at Mitch's wrist. "I am going to splint your wrist."

Mitch shook his head and walked over to the badly distorted truck lying on its side. "Forget that, we are now hiking toward Spokane."

Jake spoke determinedly. "I'll unload my pack of all but the essentials for our survival and we will leave the rest of the gear."

Janey turned to Mitch. "And, you, Professor, will be sitting still until I splint your wrist." She scrounged around in the truck's interior for the first aid kit. "Ah, there it is!" She also found a small 2x4 and rope to splint Mitch's broken wrist. Jake packed in food, water, a compass, and their map of the western U.S. Moving into action, the three of them channeled their adrenaline and shifted into a productive fight-or-flight action, in step-by-step movements.

They looked back in shock and awe. The horizon was ablaze, spewing ash and fire. They also noticed that there was a storm of gray ash raining down on them.

"The good news, Jake and Janey, is that we are probably far enough away to avoid the extreme toxicity of the ash cloud. The bad news is that the world will never be the same. Prepare for a bad spell for the next few decades."

Jake looked at the sky in awe. "Will you just look at that? It looks like it's on fire and surrounded by a halo of crimson red."

Looking up, Mitch said sadly, "Enjoy the view, my friends, there won't be too many of these awesome moments for a while. The earth is headed for a dark era."

Jake urged them on. "We may or may not be in the clear, but follow me to that shimmer in the atmosphere." As in previous doors and times within the doors, the three of them headed toward the next future of this alternate dimension of Door #6. For once, Mitch was glad to be going into another period of time. He knew his wrist wouldn't be hurting any longer.

2044

"So, where are we?" asked Mitch. "More important, what has happened to the world during our skip into this future?"

Jake replied. First of all, we are in the year 2044, twenty-four years after the blast.

Janey continued. "Secondly, of all, society is so disrupted by the eruption that it is hanging on by the skin of its teeth. Yes, the blast tore down much of our world, but the socio-economic after-effects also tears our civilization apart. The economies of nations are collapsing. Industrial-scaled agriculture is gone. The unemployment rate in the U.S. has been estimated at 50 to 60 percent. The world economy has collapsed into a super-depression. The power industries are going out of business, and individuals can no longer sustain themselves. On the bright side, wars are no longer relevant, as nations are collapsing and the scarce resources have each country focused on internal survival. Yet, there are rampant bands of rogue militias. Transportation is limited to government agencies—such as police departments—and the privileged few who are able to afford $100 per gallon at the few stations that can still stay in existence. Fully electric vehicles are on their way out since the power supply of the nation is disappearing. All within two decades! In sum, it is a cold, dark world."

The threesome landed in Miami, just outside the Miami Center for Environmental Research campus. They were in the County Police Department and facing two officers of the law, who introduced themselves as Deputy Chief Shelly Garman and Captain Ricardo. "We've been expecting you." Garman turned in her tight shiny blue uniform that seemed to move with her. She was a tall, slender woman in her thirties with wavy red hair and a strong physique.

A smile erupted from Garman's lips that only underscored her profound ability to command in her own way. "Headquarters informed us that you are from a different time and place. I don't know what the hell the chief was talking about, but I was told to Let it Be."

Mitch responded for his threesome. "Well, actually, we were at Yellowstone, the night of the eruption. Let's say that we have been living in a cave since then and we need to be briefed on what has happened since 2024."

Garmin pointed to the man next to her. "Captain Ricardo will be your guide tonight. One of your immediate questions will be how civilization could deteriorate in such a short period of time. Hold the question, as Senator Reilly will be here, tomorrow, to explain the world-wide dimensions of the crisis."

Captain Ricardo was a Latino with dreadlocks and a happy disposition. "I have been assigned to take you through the streets of Miami and Miami Beach. Won't you join me?"

Garman stepped forward in her commando boots and escorted them to the police cruiser. "Senator Reilly will give you the bigger picture later. Fair warning, though, the picture is not pretty."

Mitch, Jake, and Janey followed Captain Ricardo out of the station house. Ricardo began his tour. "This crimson haze is from the eruption—over ten years there has been a veil over the earth." The sun was setting and the skies were spectacular. The captain explained, "This is the one moment of the day that we can look forward to. They say the beauty of the sunsets is the sole joy that 'Big Girl Red,' as we call her, gave us, when Yellowstone blew her top."

Janey complained, "Why is it that whenever there is a disastrous event they name it after a female?"

Ricardo responded with a laugh. "All women are trouble! No offense, ma'am. Please, enter my cruiser." The vehicle looked like it was twenty years old and heavily fortified with steel plates and an intimidating metal cage. The car doors creaked as they opened them to get in. The driver's-side door almost fell off from the weight of the extra plating. Ricardo described their first stop as he started the turbocharged V8 engine. It was like a *Mad Max* car on steroids. The starter wound up and a cloud of smoke shot out of the bellowing exhaust. The three of them just looked at each other. "Hey, you like my new wheels? I just scored an old Borg Warner turbo engine and installed it on this thing." Ricardo was a little disappointed in the silent response, so he revved the engine and took off. "We are going to the Overtown section of the city." As they were driving, a report came over the radio of a disturbance involving rioting and looting. "This is an almost nightly event in Miami. Not to worry, though, we all stay in the car, which has bulletproof windows, and as you noticed, steel plating. We will be joined by several other units."

They raced through the dark streets, swerving through the windy sections of the road. There were no vehicles on the road or people to be seen.

When they arrived, there was a riot going on. The nighttime street was punctuated by the flames from fires at various locations. The rioters were attacking a gated community of a civil neighborhood trying to keep their lives as normal as they could. The eerie scene also encompassed the sounds of glass breaking and the discharge of firearms. Mostly looting, but there were also several fires on the street. Then they encountered the band of forty to fifty rioters. When the other five units arrived on the scene, Captain Ricardo put on his loudspeaker. "This is your only warning to clear the area. You will be gassed in three minutes if you do not disperse." Some of the crowd did, indeed, flee. The vast majority, however, stood firm and some immediately charged the police vehicles. Ricardo radioed the other units with succinct orders. "Stand ready to spray at short range and discharge at twenty feet!" Almost immediately after the orders were broadcast, the looters/rioters were approaching their vehicle and when they were nearly on top of them Ricardo pulled a lever in the front of his cruiser. The *pop* sound was followed by a discharge of crimson smoke and the attackers were immediately stunned and fell to the ground, rubbing their eyes.

"What did you spray them with?" said Mitch in an irritated voice.

"Tear gas," responded Ricardo. "Not to worry in here—our air is filtered and we have a back-up cartridge." More of the attackers kept coming forward, but were stunned when they hit the cloud of tear gas. "Idiots," said Ricardo. "You'd think they would know from all of the previous looting riots that they stand no chance of overcoming the tear gas spray."

Mitch said, "You mean that you have to spray crowds of people routinely?"

"Almost daily." Gradually, the crowd began to disperse.

"What about a using a full contingent of police as peacemakers. Haven't you tried that? I would think that it could avoid scenes like this and start to bridge the 'us vs. them' conflicts between citizens and the law."

Ricardo laughed out loud. "Where are we going to get the resources to do that? Anyway, they tried that in Philadelphia and St. Louis over the past ten years. It resulted in other rioters coming out of the deep recesses of the streets to engage the battle with guns of their own. In Philadelphia, the police casualties from this so-called non-violent attempt to restore order resulted in the death of forty-six officers just last year. In other cities, the death rate was somewhat less, but the violence is like a battle. We are in a virtual war with the looters across the nation."

"What brought us to this violent state of affairs?" asked Mitch. He had a feeling he knew the reason.

"Honestly, Doc, it is desperation. These people cannot feed themselves. The deaths from hunger across the nation in the last decade are staggering. One estimate is 500,000 per year, nationwide. Some officials estimated the homeless and destitute in urban centers at 40 percent. These disadvantaged people soon turn into looting rioters out of desperation. In rural areas the looting is less; desperate people there just quietly die. Police departments nationwide have had to cut their forces radically due to minimal budget to support our operations. We can only do this daily routine of putting out the fires to stem the violence. That is why we turn to tear gas to try to keep some semblance of order. This was only tear gas, but our arsenal includes other devices, including electrical shock radiation technology—what used to be called Tasers, but enhanced to shock large numbers of people remotely."

Mitch responded, "Wow."

As they all looked at the chaotic, dismal scene, the captain continued. "It is scary out here. The sense of order in our society is peeling away. We can only hope for the disappearance of the crimson ash cloud and a return to normal."

"I think that you may have answered my question—all of this stems from the volcanic aerosol veil around the planet—right?" asked Mitch.

Ricardo looked at him with incredulity. "Where have you been, man? I was told that you were from some other place or time—whatever that means—but don't you listen to the emergency radio broadcasts? Have you been living in a cave?" Ricardo shook his head in bewilderment. "It is the greatest socio-economic tragedy that has hit mankind—we call it the Crimson Depression—after the Yellowstone blast that has enshrouded the world with a red ash smog for ten years. We have an unemployment rate estimated at 40 percent nationwide. But the economy may be much worse in other nations. I will leave the sorry details out of this explanation. When we get back to headquarters, Senator Riley will brief you on this. For now, let me give you a tour of Miami to open your eyes." Mitch looked quizzical.

"Oh, I guess you weren't told," said Ricardo. "The arrival of the three of you from a different time and place has generated a lot of curiosity and interest. When news of otherworldly visitors reached the mayor's office, she wanted to have you checked out by a U.S. official. I will warn you that your arrival has generated a hope that you have come to us with 'otherworldly solutions' to resolve this world catastrophe."

Mitch looked back at Jake and Janey. "Do you two know about all of this?"

Jake responded, "We know of the exaggerated, almost messianic expectations, but we did not herald your coming."

Janey added, "Let it Be, Dr. Mitch. We'll explain later."

Mitch yelled out as a very close gunshot was heard and ricocheted off the side door window where he was sitting. "Okay. It's time to get the Borg Warner engine hopping again!" They could hear the turbo winding up as they spun around with tires screeching.

Mitch looked back as flames were spewing out the exhaust, leaving the crowd behind very quickly. "You popped up the boost to twenty-five! Isn't that a lot?"

Ricardo just laughed as he said, "Yeah."

They cruised the desolate streets. The entire city was one large slum. Ricardo periodically chased down gangs and threatened them with the tear gas spray. Businesses were closed and heavily sprayed with graffiti. Schools seemed to be abandoned. It was as if civilization was disappearing. Mitch looked back at Jake and Janey. "You two have been quiet as mice back there. Are you as nauseated as I am?"

Jake said, "Well, you probably guessed that we know of this dimension."

Janey followed up, "And it is depressing as hell."

After half an hour of touring, Ricardo asked Mitch if he had seen enough.

"I get the picture, but I have a few questions."

"Shoot. No pun intended!" said Ricardo with a laughing tone.

"Okay, first, all I see here throughout the city are abandoned homes and businesses, with a few buildings that are seemingly occupied by vagrants. Where do ordinary people live in Miami?"

"Tough question," responded Ricardo. "It is hard to define what 'ordinary' and 'normal' mean in these times. Our population in Miami has been reduced from 450,000 in 2024 to an estimated 40,000 in 2034. With the exception of a few isolated gated compounds, many of the residents fled to rural areas, communes and the 'new culture' enclaves. Many have died. Our culture is dying and what we have left are the survivalists and the anarchists."

Mitch considered this. "Your answer has spawned a second question. What are the 'new culture' enclaves?"

Ricardo replied, "That is a question that can be better answered by Senator Reilly."

Mitch shot his next question. "We had communes back in my home place in the 1960s and 1970s. They were socialist communities who shared possessions, responsibilities, and occupations—mostly centered on farming—and were a self-sustaining subculture of our society. Is that what these enclaves are?"

Ricardo replied, "That is the idea here, too, but the occupation—farming—is largely experimental. New crops are needed that can survive the acid rain pollution from the sulfur aerosols—and the reduced light—from the ash veil. Also, the enclaves are largely supported by government subsidies, so they are not a 'counterculture.' However, since many crops have been successful, a large part of the shared responsibility is security. The communes have been subject to attack by organized mobs and there have been two mass slaughters by these bandits, who attacked the enclaves to raid the crop stockpiles. So these communes relocated to secret areas and have created their security forces akin to armed guards. They have gated their properties and armed themselves with lethal weaponry. Important point: You will be advised tomorrow to keep this under your hat. The only reason that I know this is that I am being transferred to the Florida enclave next week. If the entire police force of Miami knew of the enclaves, all hell would break loose in chaotic demands to be transferred there."

"I need a break to absorb all of what I have seen and heard from you tonight, Captain Ricardo," said Mitch. "You have been a most informative bearer of depressing sights and information about society. Let's go back to headquarters."

Ricardo agreed that this was enough for one night, turned onto a highway, and wound out the cruiser to over 130 mph. The deep roar and dust it created was like a dream to Mitch as they sped along. He fell into a silence laced with depressing thoughts. As they turned off the highway on their way to headquarters, he shot up with renewed attention.

"Wait! Slow down, Captain. That's the Miami Center for Environmental Research that we are passing—right? I'd like to stop here—I was a professor at the college."

Ricardo replied, "Well, this location *used* to be the university. It closed down in 2025, as did virtually all centers of higher learning throughout the nation and the world. There is nothing to see here anymore—just decaying buildings." Mitch went back into his glum silence. He noticed that snow was falling lightly.

They arrived back at headquarters close to midnight where they were greeted by Deputy Chief Garman. "How did the tour go?"

Ricardo responded, "As expected, Shelly. It was quite an eye opener for Dr. Mitch. He has had it for the night."

Mitch affirmed, "Deputy Chief Garman, that was one rough tour."

Shelly resumed, "We have living quarters set up for the three of you. Get a good night's sleep. We are meeting with the senator at ten a.m." The trio was taken to rooms in the headquarters that were set up with cots and a shared bathroom. Janey looked like she was ready to fall asleep and Jake was calming himself with his music. Each had learned in their own way to deal with all the dimensions in their life of time travel. They had surely known all about what they saw tonight.

Shelly showed the twins their rooms. "Doctor, follow me. I will show you your suite set up especially for you." She guided him to his room. With her commando boots, she was practically an inch taller than Mitch. As they got to his door, she seductively kissed him on his cheek.

Mitch, on the other hand, was sapped of energy and emotionally overloaded. "Uh, I, uh goodnight." She opened his door for him and turned.

As her long lanky body swaggered away, she waved with a slow alluring "Goodnight. Maybe later, Doc."

Before he turned in, Mitch went to Jake's room and described his encounter with Deputy Chief Garmin. "So what gives, Jake? In my dimension, such behavior would be considered not only inappropriate, but grounds for getting fired."

Jake shook his head. "Think about it Dr. Mitch. This is not your dimension. The fabric of civilization is ripping apart at its seams. Think of this as a 'Twilight Zone' reality of this society that is falling apart."

Mitch returned to his room, closed the door, clicked the lock, and crashed onto the bed. He slept deeper than he ever had. In the distance you could hear the faint pops of gunfire and screams of a desperate, tormented people.

Mitch was awakened at seven a.m. by the blast of a siren. "Oh my . . . okay . . . I'm awake." He shot up like a bolt of lightning, looked around, and eventually remembered where he was. A chair held a clean change of clothes and toiletries. He showered, shaved, and dressed. When he came out of his room, he was greeted by Shelly.

"Good morning, Mitch. I was hoping to spend more time with you last night but I heard you lock your door."

Mitch responded sheepishly, "Sorry, Shelly, I was tired."

She moved on. "Did you like the blast this morning?" She smiled with a small laugh. "We sound our siren at sunrise and sunset. It gives the people of the community some comfort to know that we are still here."

"From what I saw last night, most of your residents would be happy to see that you had disappeared."

Shelly responded, "Oh, there are pockets of law-abiding citizens in the community—all gated with their own security details. It is they whom we are serving, not the rioters, looters, and vagrants. I did ask Ricardo to show you the worst portrait of our city, which is unfortunately the dominant one." She assertively put her arm under his and directed him to the stairs. "Come now, let's have breakfast before your meeting with Senator Reilly. You will get the bigger picture from him."

After breakfast, Senator James Reilly came into the conference room and greeted the trio and Shelly Garman. After introductions, he got right to the point. "Okay. I am here to brief you on the 'Volcanic Veil' crisis that is facing the planet. I understand that you are from a different time and place, Mitch, and that I am not to quiz you on your 'dimension,' as it has been described. That's fine by me. You have generated a great deal of fanfare from those who have heard the rumors, and many believe that you are a prophet or a Messiah. I must tell you that I have serious doubts about the validity of that rumor."

Mitch broke in. "I will tell you that I am most likely a stronger skeptic than you are on that point. Where the hell did that idea come from?"

Janey broke in. "We'll explain to you later."

Mitch looked irritated. "I sure hope that you two did not start such a rumor!"

Jake responded, "That's the last thing we would do. People in this dimension are desperate. However, when the rumor got started, it unfortunately snowballed and it has generated false hopes."

Senator Reilly replied, "Well, I am glad that we are all on the same page on that. So let's get started. I understand that you had a tour of downtown Miami last night and that you were shocked at what you saw."

Mitch responded, "Your world is just as devastated as one that might have been ravished by global warming."

The senator laughed. "That's the first time that I have heard that term in years. Who would have ever thought back in 2020 that the climate change they were worried about would be in the other direction?"

Mitch turned to Senator Reilly. "It is hard to fathom how society deteriorated in such a short period of time. It seems like the skin of civilization is peeling away. From my knowledge of climatology, that seems implausible. The most recent eruption of a super volcano—the major blasts that occur very infrequently—was Toba[44] and it

---

44 See https://en.wikipedia.org/wiki/Toba_catastrophe_theory.

left the human population diminished, but 10,000 people remained and civilization marched on."

Senator Reilly replied, "Back then—70,000 years ago—the world was not so interdependent. Tribes that survived the blast continued on. Today, the socio-economic interdependence of our population cannot withstand the instantaneous massive changes caused by the volcano. Unlike global warming, the effects after the Yellowstone eruption were rapid. They literally destroyed the western United States. Immediately, agriculture in the breadbasket collapsed under the ash on the ground and the crimson haze in the sky. Ash fallout occurred as far as the east coast for several months. Almost without a moment's notice, we were reduced to a nation in devastation. Anyway, what you saw here in Miami is just a microcosm of what has been occurring around the world. Particularly hard hit are cities. Jobs are virtually non-existent in almost all sectors. There is no tax base anymore. Government is stripped back to a minimum. The only agencies that are still around are defense—almost entirely focused on urban violence—as well as the climate-monitoring offices of the EPA, the Department of Agriculture's pilot project to re-seed the planet, and the treasury, which has become a joke with no revenues coming in. The largest direct hit is to the worldwide economy. The whole world is in a long-term depression. It all started with the collapse of agriculture worldwide. With the ten-month winters in the temperate zone, agriculture has shifted to the tropics. Ironically, the only nations that are above water economically are in these areas. We know very little about business worldwide. Allegedly, it is most likely focused on agricultural investments in nations like Venezuela, Uganda, and most of Southeast Asia. There may be new initiatives to support large-scale agriculture in the tropics. But since there is no network of communications worldwide, the picture of the international business world remains a mystery. The U.N. collapsed in 2025. One of the hardest things for me to see is the collapse of educational institutions. We have two remaining universities open—American University and Tennessee—with minimal enrollment and supported entirely by government subsidies. I don't know how long we can continue to do that. Many communities cannot even run high schools and middle schools. Even grammar schools are starting to close. Education is shifting more and more to home schooling."

Mitch interjected, "So when are you going to tell us the bad news?" Everyone laughed.

Senator Reilly continued, "Actually, there are glimmers of hope. One of the positives coming out of this bleak scenario is our assumption that large-scale wars are a thing of the past. Nations cannot afford to arm themselves. There is a lot of guerrilla warfare between various factions and tribes, but it is mostly every nation for itself. Did anyone tell you about the communes and the 'new culture' enclaves?"

Mitch told the senator about his conversation with Captain Ricardo last night. "Ricardo told us that you would explain more about what these enclaves are all about."

The senator responded, "Well, I'd like to take you to one of Florida's cultural enclave communities. It is located in Delray Beach, an hour north of here. Come, we'll take my car."

On the road from Miami, Mitch noticed that there was no traffic on the Florida Turnpike. "Doesn't anyone drive anymore?"

Senator Reilly responded, "It is generally prohibitive for anyone to drive. Gasoline production is so weak that the cost has risen to one hundred dollars per gallon. Once in a while you see electric cars like mine on the road. That's it. Of course, the exception to this is the cowboys on the police forces. But it is only a matter of time before we run out of gasoline and electric energy—then we will be back to horse and buggies."

When they arrived in Delray Beach, the senator pulled off to the side of the road. "Before we enter, let me tell you the conditions for visiting the community. These enclaves are projects that are highly classified. Anyone, other than the residents themselves, who enters must be quarantined after their visit, indefinitely. After the slaughters at the facilities in Atlanta and Louisville, we have to assume that the moment that people know about where they are, the exposed enclave will be attacked. The thug militias, nationwide, are getting stronger and bolder. However, the president has decided to give you an option. If you return to wherever you came from after the visit, you will be allowed to do so as long as you never return to our time and place. We all have a feeling that you are headed out anyway, so we will not stop you."

Mitch interjected with a question. "If everyone is so convinced that we have come to resolve this emergency, why would you allow us to leave?"

The senator replied, "Most people think that you are literally going to eliminate the crimson cloud from the atmosphere. Many of us in Congress and the president's inner circle have a different explanation. You are traveling through different times and different dimensions. By seeing our plight, you may be able to influence conditions of the past and allow society to prepare to respond to the Volcano Climate Emergency. The president has a science advisor named Reverend Tyndall, who is very credible, very trusted and seems to have insights into where and when you came from and when and where you are traveling to. One other advisory before you make your decision of whether to stay or leave here. The pressures on the three of you from a desperate population of suffering people will become so great that you may not be safe here. You also have the option of not visiting the enclave and staying here. Consider whether it is worth seeing. Bear in mind, if you do decide to visit the enclave, the full picture of our society will become clear to you."

Mitch turned around to talk it over with Jake and Janey. The three of them agreed on an exit strategy. He responded for the trio. "We will leave here after the visit." Let's see this beacon of hope, and then we will move on to a different dimension."

The senator smiled. "I hoped that you would see it that way."

They entered the grounds of what appeared to be a correctional facility through a gate. There was a concrete wall all around the community, topped with barbed wire. It was a virtual rampart protecting the facility. The two security guards were outfitted with weapons. The initial buildings they passed certainly looked like a prison. Maybe it was. The senator then drove them down a dirt road and into the forest beyond the buildings. After a mile of driving on a very bumpy road, they came to another gate and another rampart perimeter to the site. Senator Riley again showed his pass, and the security guards—noting that the senator had three authorized guests—opened the gate. When they entered the grounds, Jake, Janey, and Mitch were entranced by a very large domed facility made of glass, which dominated the area. The greenhouse had windows that were not quite clear. They seemed to be glimmering with a luminescent material.

"Now you see why we have to protect this facility? Do you understand why the first attempts to integrate these enclaves with the outside world resulted in attacks and slaughter of the residents? What you have called security guards are actually U.S. Army troops on permanent protection duty. This is one of the select, federally supported domestic programs. As you will see, this represents our last effort to adapt to the crimson veil in the U.S."

They entered the greenhouse and the Senator ushered them into a room with comfortable chairs, wall hangings, and large windows that flooded the area with light. On a long mahogany conference table sat five pitchers of water, each tinted a different color.

A woman and a man entered the greenhouse. The woman was dressed in a long flowing satiny dress. She had wavy blond hair, sad brown eyes, and a sense of glowing beauty that enshrouded her face. The man was dressed in a brown robe with a hood. He had dark curly hair and a full beard. "Welcome, senator and guests. I am Tobias. I am the superintendent of the Sunshine Enclave. This is my wife, Carlisle. I will be your host on this tour."

Senator Reilly, Mitch, Jake, and Janey introduced themselves.

Tobias asked, "How much have you briefed them, Senator Reilly?"

The senator replied, "They know that this is a top-secret community and that farming is your main way of life."

"Well, I understand that you are all from a different time and place. Do you know about the big blast of Yellowstone in 2020?"

Mitch responded, "We were there at the eruption, but we came into your world not knowing about events between then and now."

Tobias' eyes widened. "Wow! You were there? You will have to share your experience later. For now, however, let me fill you in on the decade since the eruption. Then we will take a tour of our facility."

Tobias took a sip from one of the pitchers, whose water was a shade of peach. He offered some to his guests. "The water has been infused with a fruit and vegetable elixir and will give you a calming effect." Mitch filled a glass with water that had a yellow hue, labeled "lemon." Jake and Janey filled theirs from the pitchers with red water labeled "vegetable medley." Almost immediately, the three of them felt a broad calm wash over them.

"Don't worry about the effect; it is not an untested chemical drug. Our beverages have been extracted purely from our homegrown produce and our chemists have found a way to extract the right combinations of these fruits and vegetables to stimulate a sense of calmness—like tea. It is purely natural."

"No complaints here," said Mitch. They all laughed.

Tobias spoke. "So, the effect of the volcano on the planet was immediate and severe. Very few people within one hundred and fifty miles of the blast survived due to the suffocating toxic gas plume."

Mitch thought to himself. *We made it to Idaho; we were about 250 miles out—close call.*

Tobias continued. "Our satellites monitored the eruption, which continued for days. Yellowstone National Park was transformed into a pit deeper than the caldera that was there before and extended to a hundred-mile radius. The volcanic veil that enshrouded the entire planet was a deep shade of crimson. It was quite beautiful in its appearance, especially at sunset, but quite deadly to our planet. The devastating fallout from the ash was felt throughout the western states of North America. Nevertheless, a smaller fallout of ash was evident throughout the planet. This was unprecedented. The blast from Yellowstone reached so high into the stratosphere that it caused a worldwide cloud that persisted for several years.[45] Even Europe reported ash. Indoors—both homes and businesses—were laden with ash from air intrusion from ventilation systems and ash that seeped into buildings from windows and doors. So much of our infrastructure was damaged or destroyed. Roads were pockmarked from the acidic ash fallout that ate away at the asphalt. The equipment of manufacturing facilities was severely damaged or destroyed. Obviously, the closer to the volcano, the more severe were the effects. An area three hundred miles around the super volcano was uninhabitable. Casualties in the western U.S. and Canada were estimated to be in the hundreds of thousands. Not just from the toxic gases but from the spontaneous fires that destroyed buildings and their inhabitants. There was a mass exodus of homeless survivors from the volcanic plume zone migrating eastward or southward. The prevailing westerly winds spared California and other coastal areas from the devastating effects. Many people scattered in all directions to set up homes. There were large tented villages established around urban areas receiving hordes of migrants. The survivor tent villages were created and supported by the federal recovery efforts, which rapidly drained the budget, creating gargantuan deficits. And then, there were the aerosols. Unlike ash, these gases did not settle out over one decade but remain with us

---

45 See http://volcano.oregonstate.edu/how-high-can-explosive-eruptions-go-and-how-far-can-debris-and-ash-be-spread

still, blocking our sun and keeping our temperatures low. The aerosol cloud encompassed the entire earth."[46]

All were silently listening.

The senator continued after a brief moment to sip his water. "Fortunately, by 2026, the ash fallout ceased. However, the crimson veil that enshrouded Earth remained. The aerosols, which were fine particles suspended in gases, remain in the stratosphere and atmosphere, descending ever so slowly toward the earth. The ash cloud was washed with every major precipitation event. But, the aerosol cloud—largely composed of sulphates—reacts with precipitation to create acid rain. I am told that you are a geologist, Mitch, so you know about the effects of acid rain."

"Yes, Tobias, acid rain is mild enough so that people are not affected by acute burns but many lakes in the late 1980s lost fish due to a similar effect."

Tobias continued. "Worst hit was agriculture, which virtually collapsed in North America and eventually shut down in the entire temperate zone of the northern hemisphere. We have heard rumors about the fate of the rest of the world, which allegedly experienced similar devastation. Radios broadcasts, satellite communications, and computer networks were eliminated due to the worldwide veil of ash. Airline travel and even shipping were unthinkable due to the ash veil that would destroy their turbines and severely reduce visibility. So, the U.S. turned inward."

The senator paused and took another sip of his drink. "Meanwhile, our earth's climate seemed to shift into a different mode. We had seven years of nuclear winter from the ash fallout. Even after the ash fell out, however, we remain in persistent cold and dark. You experienced this summer's snowfall in Miami. Canada and the northern U.S. have become wastelands due to extreme winters that last the entire year. Gradually, the northern populations migrated south and our cities from Washington to the south coast have become virtual refugee camps. As you saw in Miami, order has broken down. Most people are now what were once considered low income, and there has emerged a lower class that has been cruelly named the 'volcanic dregs.' "

The room was very silent and Mitch was mesmerized by the description of planet earth. "So when are you going to give us the bad news?" The tense atmosphere broke.

"Oh, I'm getting there," said Tobias." Let's start with the geological estimates of the blast. The eruption of Yellowstone was estimated to be a middle-class event in the scale of severity of the historical eruptions of super-volcanoes that previously wiped out most species on Earth. So . . . it could have been worse. There have been some estimates that the aerosol fallout will start to settle down a decade from now. Other estimates crank that up to one hundred years. Of course, that will not help us in the lives of my

---

[46] See https://www.geolsoc.org.uk/~/media/shared/documents/education%20and%20careers/Super_eruptions.pdf?la under the section entitled "Hazardous effects of Super Eruptions.

generation or our children's." Tobias took another drink of his peach beverage and continued. "However, our best hope to try to adapt to these harsh conditions that we inherited from Yellowstone is centered on our 'cultural enclaves.' You know how top secret these communities are, though most people outside do know we exist. They just don't know *where*. Anyone who wanders close enough to recognize our community is detected by our military security force, arrested, and brought to a federal detention facility that is, in essence, life imprisonment."

The senator broke in. "The dominant function of our armed forces provides security for these communities. This one alone has hundreds of army and special unit personnel guarding 1,875 enclave citizens. Certainly it is enough to repel a typical guerrilla attack, but equally important is to guard its secrecy." Mitch was fully absorbed in the conversation. Jake and Janey looked somewhat bored, as if they'd heard this all before or instinctively knew the story of this 'dimension.' Almost unnoticed was Tobias' wife, Carlisle, who sat at the other end of the table sipping her green drink. She looked as if the burdens of the whole world were on her shoulders.

Tobias continued. "Currently there are about 1,000 of these enclaves throughout the nation and there are plans to double the number over ten years. Our hope is to seed the nation with these cultural centers of social evolution. Sadly, we have given up on society's ability to stop the erosion of civilization outside these walls."

The senator added, "We have reason to believe that Europe—and perhaps elsewhere—has established their own enclaves. Worldwide, the communications between enclaves here—and perhaps in other nations—may become the sole basis of our international relations, but for now, we do not communicate beyond the enclaves in the Southeast U.S. We will eventually reach out and coordinate an information-sharing network with other nations who have these enclaves. We see this effort as 'seeds' of a cultural renaissance. Each community operates differently. We expect growth, but we also expect that some of these communities will fizzle out—survival of the fittest. Let's take a break and check the weather."

They went outside, it was now snowing lightly. Mitch brushed flakes off his face.

The Senator noted the weather. "We'll have to wrap this up quickly . . . the storm is approaching. Let's go back so I can finish my briefing."

Back inside, Mitch asked: "It's like you have two worlds. There are these utopian villages and the chaos outside of these gates."

The senator looked very serious. "That's the way it is. We expect that the world outside of the enclaves will die out in time."

"Wow!" said Mitch. "Yet, with this severe depression, how is the federal government able to subsidize these enclaves?"

Senator Reilly responded, "Remember these are self-contained communities. We do provide subsidies, but they are also self-contained economically—akin to autonomous states or even nations. Their economy is supported by the success of agriculture, and we are hoping that success will lead to new business and commerce. In fact, other enclaves have other commercial and industrial specializations. Our new 'farm fields' are huge greenhouses, protected by glass enclosures that are not affected by acid rain and are supplemented with artificial light. Energy is provided by a large windmill farm. Our big government secret is that instead of having fifty states, we now have 1,000 enclaves. Actually, some of these enclaves have a thriving economy, with no deficits, no debts."

Mitch looked perplexed. "But who decides who lives here?"

Finally, Carlisle looked up and spoke. "That burden is on me and similar 'selectors' at each enclave. We research the talent and fitness of people outside of our walls and select those who can best contribute to the sustainability of our enclave. It is a severe ethical burden that I carry, but you must understand that this is our planet's last hope for living this era out until the crimson volcanic veil is lifted."

The room turned quiet for a long and uncomfortable period. Finally, Mitch broke the spell. "So this seems like quite a massive operation. All centered on this one greenhouse?"

The senator laughed, as Tobias responded. "Come, let's take you for a tour of our community. You are only seeing the tip of the iceberg." Tobias took them through the football field-sized greenhouse. It was filled with row after row of vegetables and fruits. He explained that the panes of windows were impregnated with solar-enhancing chips, which magnified the sun to increase the incident solar radiation to a suitable range for plant growth. The solar chips' magnifying ability was self-monitoring and self-adjusting to allow maximum growth and sustainability for the crops. There were greens and legumes, fruits and vegetables of every kind. There were even orange trees. Tobias explained that this particular glasshouse was a "community garden," a place where the citizens of the enclave could have their own private growing area.

"This place is huge!" said Mitch. "What is your population anyway?"

Tobias responded, "We are 1,075 and growing at a controlled and manageable rate. We have a staff of ecologists, agronomists, engineers, and other disciplines that determine the optimum sustainability rate and we recruit highly specialized and talented professionals on an as-needed basis. We also have a large family planning initiative to determine fertility limits of our residents of child-bearing age."

Mitch shook his head. "I don't even want to know about how you enforce the size of your village population here."

Tobias responded, "These are tough times that require hard decisions." After the greenhouse tour, Tobias took them outside, where it was still snowing. "They are predicting a blizzard for tonight. We want to give you a full tour, but we will need to get you to sleeping quarters no later than four PM. We cannot risk visitors getting lost in a storm. Besides, you do not look adequately dressed for our weather here."

For now, the weather was pleasant. The air was brisk and the gloves, jacket, and wool hats Carlisle had provided them were comfortable. It is so weird that there was a need to be preparing for a blizzard in south Florida in July. They enjoyed a pleasant walk through the woods to their next destination. Mitch then noticed that they were approaching a clearing. When they came out of the forest, they were in a very large meadow—Tobias said that it was 500 acres.

But the amazing aspect of the clearing was the greenhouses and other buildings. The surreal scene of large industrial-type complexes overwhelmed Mitch. The landscape was covered with mega-sized greenhouses, barns, office buildings, and a complex that looked like a water or sewage treatment plant. All of these buildings were woven in an idyllic setting with fountains, walkways, and the energy of people moving around from building to building doing their work. Beyond the complex, there was a very large meadow that seemed to be dotted with wildlife; all of this supported by the windmills and the domed greenhouses the size of football fields. Mitch turned to Jake and Janey. "With the exception of the crimson veil in the atmosphere, this is unreal in contrast to what we saw last night. It is like a chrysalis waiting for the sun to reappear, allowing it to fully hatch."

Janey smiled, "Do you think?"

Jake observed, "Remember there are 1,000 of the enclaves throughout the nation. This is the future of civilization in this dimension of time, if there is a future."

"What do you think?" asked Tobias.

"Overwhelming!" said Mitch. "This is a reverse image of what exists outside of your gates."

Tobias ushered them toward the first building. "This is our agri-energy-manufacturing complex. It is the heart of goods and supplies needed to support our enclave and the basis of our economy here. Let's take a brief tour of our industrial compound. I want to move us along so that you can see the residential area of our enclave." Tobias took them through each building. There were more agriculture production greenhouses. There were manufacturing buildings to produce the goods needed by the residents. There were barns with farm animals of every sort. There were research buildings for environmental monitoring with several initiatives of sustainability modeling for the enclaves. There were gymnasiums. There was even a museum. They spoke with many residents—a stark contrast to Miami where most people were holed up, nowhere to be seen.

By the end of two hours, Mitch's head was spinning. "I need a break, Tobias. All of this is very interesting, but I am mentally exhausted."

Tobias replied, "Yes, it is time for lunch and a nap." They walked outside again and Mitch noticed that the wildlife on the meadows were farm animals—cattle, horses, and other domestic livestock. On the ridge above them was the windmill farm. Tobias quickly ushered the group to a building with dining halls and cafeterias. They had lunch and discussed what they had seen. After lunch, they went into a very large room that was sectioned into cubbies. "We have adopted a custom for a two-hour lunch and siesta from noon to around two. We find that this habit encourages creativity and productivity in the afternoons. The group then took a twenty-minute nap in a quiet room, equipped with scores of hammocks. Afterward, Mitch did, indeed, feel refreshed.

Tobias smiled. "Feel ready for the second half of our tour?"

Mitch smiled, too. "I sure am! Let's see what else you have in this utopian village." He noticed that Jake, Janey, and Carlisle were still very quiet—and had been all day. The smile left his face. Mitch felt disconcerted about the remainder of civilization withering outside of the enclaves.

Tobias took them to a separate pathway—this one paved—through another section of forestland. After about an hour, they came to another clearing. "This is the second area of our community: the residential complex." They were looking at a city, which actually looked normal. People were traveling around on electric golf carts. There were many supermarkets, several neighborhoods of housing, restaurants, stores, and even banks. It reminded Mitch of Disneyworld. Everything was so new, so clean. They walked through the city, talked to people, and the stroll in the light snow was very pleasant. Tobias took everyone to an "automat"—a food service modeled after the 1950s restaurants in New York City—which charmed Mitch. As they were finishing their dinner, the blizzard started in earnest. Tobias looked at the gathering storm and prompted them along. "We need to get back to the community garden greenhouse to end our tour. We will use an electric cart . . . better yet, let's use horses." They walked to a stable down the street, saddled up, and the five of them were off. Mitch found the ride exhilarating as the horses raced them through the blizzard down the path to the greenhouse where they had started their tour.

In the conference room, they warmed over coffee and dessert. Then, Senator Reilly opened up a conversation about their parting. "So, my friends—the choice is yours to move on to where you go from here. I know that we discussed this previously, but I put these choices on the table again in case your tour here has changed your mind. We will be happy to have you stay with us, but it would be in permanent detention in this enclave. Alternatively, we could place you in detention off site. The center we would take you to is not a prison, it is more like a commune and you would be comfortable. However, you would be monitored for the remainder of your stay here and isolated from the outside world. Do not take this personally, but we can't trust anyone who has seen the location of our cultural enclaves. The final option that you have is to return to your different dimension beyond this dismal world. I suspect that you will take this alternative."

Mitch looked at Jake and Janey. The three of them nodded. Mitch spoke. "You have been very kind hosts and have given us an eye-opening picture of what your world is like."

Jake added, "We know that it is your hope that we will be able to use the tour of south Florida that you have given us in a manner that will help resolve the crimson veil that enshrouds you. I cannot give you any guarantees that we will. However, our journey to your dimension has a higher purpose. We cannot describe that purpose to you. Even we are not fully aware of this But as we continue our journeys to different 'dimensions,' I can assure you that your world will be in our hearts."

Janey followed, "And now, we must go."

Mitch paused and reflected on what he had experienced. The same crimson bubble appeared before Jake, and Janey, and himself. They jumped in. As the bubble ascended above the greenhouse dining room, they waved at their hosts. "I feel somewhat sad leaving them here," said Mitch. "Their world is so devastated."

Janey replied, "Dr. Mitch, this is all they know. Their memories of the pre-eruption years have receded and they are dealing with the crimson volcano veil as if it had been with them all of their lives."

Jake added, "If you look at it another way, they are in an exciting race to save civilization through the enclave world that they have created."

Mitch looked a bit melancholy. "I guess we will never know."

Janey patted Mitch on the shoulder. "You forget that we are still on our journey. We are headed to a new 'era' through this same crimson door."

Jake said, "Dr. Mitch, we will have some time before we arrive there. It is a good time for us to reflect on what we saw in the post-eruption world. What do you think?"

Mitch scratched his face, again forgetting that he no longer had his gray beard. "My first reaction is how this eco-disaster is a mirror image of what we saw through Door #1. The climate change that the world experienced for neglecting to reduce our greenhouse gas emissions resulted in the same degree of planetary devastation, but in a different direction. The factor that gave them a ruined planet was the same as in this crimson world: the suddenness of the flip in atmospheric conditions. In the case of the volcano, it took one big blast at a single moment of time. However, in the case of global warming, it took two centuries of industrialization and emission of greenhouse gasses. In the short term, the volcano world was more severely devastated due to its instantaneous change of the atmospheric climate conditions."

Mitch continued. "However, if you think about it, the global warming conditions will return in a relatively short period of time, geologically. Two hundred years is a micro blip in the tens of thousands of years that it takes the climate to naturally flip into an ice age and back.[47] The veil of this crimson world will eventually be removed, when the aerosols and ash are washed out by precipitation. It may well be that after the sun-reflecting veil lifts, conditions will return to pre-eruption conditions and CO_2 levels will still be elevated

---

47   See https://en.wikipedia.org/wiki/Interglacial

above the level of the early 1800s.[48] It takes decades to centuries for the CO_2 levels to stabilize or decline after emissions stop."

Jake and Janey looked at each other. "Very profound, Dr. Mitch," she said. "Now let's hear about your personal observations pertaining to our journey. Let's talk about the questions that you raised while we were in Florida."

"Okay," said Mitch, "I have no doubt that we are headed to a future period of this world—a century or so forward based on our journeys through the other doors. I will not conjecture what we will see there. I will Let it Be what it will be. But yes, I do have questions. First of all, I did not feel like I was in an active role in this world as I was in previous journeys—for instance, in our struggles dealing with geo-engineering. Here I felt that I was just an observer. What gives?"

Jake responded, "Dr. Mitch, you *were* just an observer here. Sometimes it is important for you to be 'in' these realities, other times you need to just see and feel the world of a different climate scenario."

"Well that raises a broader question that I am a bit reluctant to ask," said Mitch. "What is the big picture? How am I to use this information for my own world?"

Jake responded, "I don't think that you are ready to deal with that question yet. Remember we are on this journey with you as your guides. So, for now, just Let it Be."

Mitch considered this. "Okay, I'll Let it Be, but what about the role of both of you? Do you know all? It seems to me most of the time that nothing of what we observe is a surprise to you. Other times, you do seem to be taking what we see in a less omniscient way. For instance, I did sense your fear when the Yellowstone volcano erupted under our feet."

Janey replied, "It is the same with us as it is for you. Sometimes we are just observers. Other times, we are part of the here and now of the 'dimensions' that we visit. However, in our case, we have a general perspective of a broader view of all of the dimensions that we visit. Let's leave it at that."

"Are you human or something else?" asked Mitch.

They both responded in unison, "Let it Be."

2162

At that moment, the crimson of the bubble that they were floating in seemed to have a much lighter hue. It appeared that it was almost normal white light, with a just a tinge of the crimson. Then the bubble broke and they were back in what appeared to be the greenhouse of the Delray Beach enclave. There were two people sitting at the conference table, one in a satin gown and the other in a brown robe. Mitch was surprised. "What? It's

---

48 See https://www3.epa.gov/climatechange/science/future.html

Tobias and Carlisle. We are back in the same place! Or maybe we have only advanced a few years—is that right?"

Jake responded, "No, Mitch, we are in 2162. Take a closer look."

The man coming toward them was not Tobias. Mitch then noticed that the woman was not Carlisle, either. He closely observed them as his thoughts assessed the two people. *They have a peaceful sense to them I have not seen before. This is intriguing indeed! I wonder if this is just the way of this new world. What happened to the rest of the outside populace? Wow, when they look at me I feel so different. As if . . .*

"Welcome. I am Terraldi Ter. I am the current superintendent. You can simply call me Ter. Meet my associate, our ethicist Malby. We understand that you are from another time and place. I have sensed from our 'stream' that you are here in the observer role for a while and then you will jump into an active role down the line."

Jake whispered to Mitch, "Short for 'stream of consciousness.'"

Janey added, quietly, "We'll explain later, but note that Ter has more of an intuitive sense of your journey to this age and the previous places where you have visited. Still, be careful to avoid talking to him in detail about the other worlds we have been to."

The three travelers introduced themselves to Ter and Malby. Mitch looked around and saw that it was not the same greenhouse after all. First, it was much larger. The size of three football fields, and the top had a retractable roof. He also noticed that the sky had lost its crimson hue. In fact, the sky was a sweet blue. Mitch smiled at this and asked if the retractable roof meant that Florida was finally clear enough and warm enough to allow crop growth.

Malby responded, "Well, we are not in what you call Florida. We are 600 miles north in what used to be Georgie."

Mitch smiled. "You mean Georgia."

"Yes," she responded. "Let's sit and Ter will fill you in."

Mitch exclaimed, "This chair is amazing! Or is it a pod?"

Malby smiled. "It conforms to the needs of your body."

"So, you are from an earlier period? How long ago?" Ter asked.

Jake responded, "We just skipped over from 2032, when the enclaves were first being established. Do you still call these cultural enclaves?"

Malby informed them that their community was named Harmony.

Mitch asked if they knew about the Delray Beach enclave.

Ter replied, "We had contact with them until about one hundred years ago before the breaking-up period. Before you ask additional questions, let me give you the full history of what we inherited from the earlier age, which leads up to where we are going in a few days."

So Ter told them the history of the last 140 years. "The experimentation period of the enclaves was successful. Nearly all of the communities succeeded and thrived. However, the 'utopian' atmosphere of the communities—such as the Delray Beach settlement you visited—faded into more typical societies: more normal settings, differentiation of economic classes, poverty, crime, economic downturns and the politics of these places made them more similar to the world of pre-eruption years. The transformation of the utopias to real human habitations was catalyzed by the withdrawal of the support of the U.S. government. We had to make it on our own."

Mitch said, "I doubted that the funding would last long. Does the federal government give you any support at this time, such as technical assistance for engineering or science?"

Malby and Ter looked at each other. Malby responded, "The U.S. government is no more. It collapsed in the middle of the last century. We are now an independent federation. The federal security forces provided to us by the former U.S. government have evolved into limited armies for each of the 'enclaves,' as you call them. We have trained succeeding generations into an armed force of eighty guards for our location. Our knowledge about the world beyond our North American union of enclaves is sketchy, at best."

Mitch responded, "Wow! Eighty guards seem rather scant. What do you do about security?"

"Interesting question, Dr. Mitch," replied Ter. "That question leads us to our current initiative. So, before I tell you about where we are headed, let's get back to our history, and I will summarize it concisely. I can skip the details of the evolution of our 'seeded enclaves' into real living communities. Fortunately, the relinquishing of support started gradually. It was tough on us to start setting up our own economies without outside aid, but we adapted. The government also relinquished their military support gradually, and we trained our own military personnel for an army. However, the collapse of the U.S. occurred rather suddenly. In addition to going from little support to no support, we also lost communications with the outside world. We have had to live here on our own without knowing what is happening out there."

Mitch interjected, "Why did you abstain from venturing outside of the gates? Is it the roving bands of gangsters that we saw in 2032? With an army of your own, why couldn't you just foray out in explorations?"

Ter responded, "Not that simple, Mitch. Before the end of the U.S., we were, of course, briefed on what was going on out there. As you know, we were in secret locations. We still are. The so-called 'gangsters' out there grew into guerilla militias that were quite powerful. A coalition of militias struck U.S. weapon stockpiles in a coordinated campaign in 2035, which gave them tremendous military capabilities. Many of the former cities of the U.S. were easily taken over. That is when the U.S. began withdrawing from our communities. The militia armies were just too much for the former nation to deal with. Outside, civilization was crumbling. So, you can see, we were relying heavily on our anonymity to stay under the radar of these militia wars. We heard that many of the enclaves were discovered by militias that mounted a siege campaign to take them over. In two instances, the militias did prevail, resulting in mass slaughter and dictatorial takeovers of the communities. However, in most cases, the enclaves held strong and repulsed the attacks. Fortunately, around the time of the U.S. collapse, we learned that the militias were disappearing. At the same time, civilization outside of our walls had crumbled. In 2055, we were briefed that the U.S. population had diminished to an estimated one million people who were distributed in scattered tribes, the gangster militias and, of course, cultural enclaves. Most people were dying from hunger or exposure to the cold. By 2060, we learned that the militia attacks had all but ceased, but that there were still bands of murderers out there with advanced weapons. Before the U.S. collapse, in 2065, we were advised to stay hidden for seventy-five years. They thought that if we were not discovered by then, the militias were likely to be gone from the earth or at least from our continent. One of the big hits for us was the loss of communications and trade with other enclaves. We had to quickly stock ourselves with enough supplies and armaments to protect ourselves and the lack of electronics caused us a big setback, but by the 2070s we did manage to find our own independent way."

"So, when are you going to tell us the bad news?" quipped Mitch. Ter and Malby laughed while Jake and Janey had heard this quip too many times for it to be funny and just shook their heads.

"Well, there is good news . . . all about the climate and the crimson volcano veil, which started lifting. Climate scientists in some of our enclaves developed monitoring devices and the CO_2 levels oddly stayed high. In fact, they were slightly higher, at 425 parts per million, compared to pre-eruption years.[49] At the same time that the veil was disappearing, our average annual temperatures started to rise toward pre-volcano levels. As the crimson cloud lifted, each of our enclaves started to thrive again with specializations of their own."

"Wait a minute! You told us that each enclave isolated itself from the outside world. How could you know this and how was this possible? How did you obtain raw materials for manufacturing the commodities that you need to survive? How could you survive without some level of trading commodities? You have an expansive area for one community, but it is hard to fathom how you could exist totally isolated from the outside world. Even hunter-gatherers had the whole world around them to seek their basic necessities. You seem to have evolved beyond a primitive stage of cultural evolution, during a period when you were totally isolated. How is that possible?"

---

49 Carbon dioxide does not wash out with the ash and aerosol. Even if carbon dioxide emissions cease (such as the time during the non-industrial period following the Yellowstone blast), the carbon dioxide emitted to the atmosphere from the nineteenth and twentieth centuries will remain for thousands of years. Hence when the sun blocking veil of aerosols lifts, it is reasonable to assume that the global warming of the early twentieth century will resume.

"You are a quick study, Mitch. Yes, I skipped over something important here," Ter replied. "Yes, we were advised in 2060 to stay inside our enclave bounds. I believe that the U.S. in their demise may have been very concerned and overly conservative about our safety."

"Anyway, we tried total isolation for a few years and everything that you just said started us on a road to decline. Hence, in the late 2060s, we started venturing outside of our gates and scouting reports indicated an absence of the militias. So, around 2070, we set out to search for other enclaves. We found many of them on the east coast that still existed, some thriving. We had a unique society, different from all other enclaves. We shared and traded with each other. Diversity is stability, after all. We did not see roaming militias or bands of bandits. Nevertheless, we stayed cautious and kept our citizens within our gates and periodically dispatched traders and diplomats with horse-drawn wagons to these other communities in the east. I am proud to say that we had a breakthrough in technology here at Harmony in 2075: solar cell-powered hovercraft. We established a network of trade centers in the east and created the concept of a compact among communities and agreed to an American summit of enclaves in 2164. That is this year. That is where *you* come in, Mitch."

Oh no. Mitch asked Ter what he had in mind.

Ter responded, "I know that you are aware of your observer versus active participant role in the dimensions that you have visited and Jake told me that you have an impressive scientific background in geology and climatology dating back to pre-volcano times. We want to put your knowledge to use in guiding us, Mitch. As part of our summit, we have established a science and technology initiative aimed at guiding us into the future of our growing communities. We want you to participate as the lead climate scientist. No one really knows what the future will bring to us. After 150 years of darkness, we are finally coming out of the tunnel. We want to know what to expect in the future."

Mitch thought about this. "I would be honored to serve, sir, and I understand that I will be flipping roles from being an observer to a participant."

Jake and Janey looked very pleased and bumped elbows cheerfully. Janey whispered to Jake, "Dr. Mitch has really evolved. In his path to enlightenment, he can really contribute here."

Jake whispered back, "You bet, sister, we have done our jobs well in guiding him along."

"So, when is the summit?" asked Mitch.

Ter replied, "Next week. We leave tomorrow."

"Whoa!" said Mitch. "How am I to get oriented toward the particulars of this climate and the ideas of how society is going to monitor and manage climate?"

Ter smiled. "You forget that when you are in the participant role, you intuitively gain the knowledge of this time era. Also, we will be arriving three days early. This will give you time to meet the other participants of your panel and get up to speed."

Mitch considered this. "Okay. Where is this summit?"

Ter responded, "We will be traveling to an enclave in the area that was once Pennsylvania. You will get to travel in our solar cell vehicles—one of the luxury models." Then, beyond the 400 mile range of these vehicles we will meet up with a group from the Richmond, Virginia enclave who will escort us the rest of the way."

"What about the militias?" Mitch asked. "How sure are we that they have died out?" Ter gave Mitch a friendly slap on the back. "That is not a worry anymore. Most of the enclaves abandoned their gates decades ago. No attacks. No intrusions. However, in Harmony, we are staying cautious, since the last reports of massacres from militias were here in this area of our continent. "

Mitch shook his head. "What do we know about enclaves in other areas of the former U.S. and the rest of the world?"

Ter replied, "Our east coast network made forays to other areas of the nation and established contact with their networks. They will be sending their own representatives to the Summit. As far as the rest of the world, who knows? Maybe we will set out to Europe soon to track them down. However, we have no contact with surviving civilizations elsewhere."

Mitch asked one more question. "How were you able to find these other enclaves?"

Ter again replied, "Before the U.S. died out, their technical support team trained us in their GPS technology and gave us exact coordinates of each enclave in the country. They also gave us written directions and maps."

The trip to Pennsylvania took seven days and was uneventful. The roads were, for the most part, overgrown with underbrush but surprisingly intact. They had a few encounters with individuals or groups who appeared to be hunting and a few vagrants begging for food, but there were no gangs, militias, or individuals who appeared hostile, threatening or dangerous. They stayed overnight at the northern Virginia enclave. They learned more about civilization outside of the Harmony gates, including rumors about the disappearance of Washington, D.C., coinciding with the time when the U.S. went lights-out in 2065. By the afternoon of the next day, they approached the enclave in Pennsylvania called Steel Town. As they crested a hill five miles from the location, it appeared: glaring at them were tall buildings and smokestacks. The stacks were belching yellow, black, and white gases. When they arrived in the enclave, they were greeted by a sign that said, "Welcome to Steel Town, the manufacturing enclave."

Oh no. Is civilization back to the same dead-end track we were on in 2020? Mitch turned to Ter. "No secrecy here?" Ter replied, "We heard that Steel Town came out of their closet in the early 2070s, but I am frankly stunned myself at what we see."

As they drove through the city, Mitch noted several vehicles and the streets were filled with people going about their business in the same rushed manner as the New Yorkers and Philadelphians of the pre-eruption years. *Wait a minute; this is western Pennsylvania, where Pittsburgh used to be. This enclave must have taken over the old Steel City of the past.*

A few minutes later, they entered a gated compound somewhat removed from the city. Ter looked around and said, "This appears to be the original hidden enclave. It must now be the city's administrative center."

After they entered the security gate, they were taken out of their vehicles, their bags inspected, and each of them was interrogated before they could proceed. Then, they entered a lavish building marked Government Offices and led up to a room that was marked "President of Steel Town." Entering the room, they were greeted by three men.

A tall, slim man with a handlebar mustache approached them with open arms. "Welcome, welcome, Harmony people! I am President Prelee. These are my assistants, Gutowski and Heinz." Prelee took them into a conference room where they were offered wine, water, or elixir drinks that were in pitchers across the table. Mitch was amused that such elixir drinks were still around. He indulged in the strawberry flavor.

Prelee gave them a briefing on the summit, which would start in two days. "We have 312 eastern enclaves that will be represented. There are, to our knowledge, no others that are present in the nation. A team from Mexico heard about our summit and arrived yesterday. Apparently, Mexico has become an agricultural breadbasket and is our only link to the outside world."

"What about Canada?" asked Mitch.

"It's a wasteland. There are most likely no human habitations and perhaps no human survivors. We also assume that the west is gone."

"Wait a minute!" said Mitch. "I'm confused. I was told in the south Florida enclave some time ago that there were 1,000 enclaves in existence. Why are they not represented at the summit?"

Ter responded, "Mitch, I know that I told you that. But I also told you that many of the enclaves died out. Prelee continued. "We are all playing catch-up as our communities are emerging out of the dark. The U.S. had planned thousands of enclaves, but only seeded 1,025 before the collapse of the nation. Of these, 70 percent collapsed. We are the survivors. There may be more that we are not aware of, but our surveys over the last ten years have been very extensive. In short, we are the cream of the crop, the fittest that survived. It is we who will carry civilization forward. Thanks to the foresight of the former U.S., we are endowed with the wealth of scientific, cultural, and political leadership of those of us who were selected and survived. We are well poised to rebuild all that we lost.

"Mitch, we now know that two enclaves—the organic growth community in Kentucky and here in Steel Town, have done more than survive. They have thrived. They are on track toward rebuilding the society that we lost from the Yellowstone eruption. They have re-established manufacturing, oil production, and agriculture on a large scale. Their economies are flourishing."

Prelee further explained. "Mitch, we have a plan that was developed between our community here in Steel Town and the Blue Grass enclave in Kentucky. We plan to stoke up productive growth and industrial expansion in the next twenty-five years. Society was down and out for more than a century, but our knowledge is still with us. We are prepared to launch a second industrial revolution. And that is where you come in. With your scientific knowledge in geology and climatology, we want your input in guiding our world to our planned exponential expansion of growth in a manner that respects the environment and avoids future eco-disasters that may threaten to bring down society. From the start, we would like to position ourselves to the period of the twentieth century, where technology controls pollution from the start. That is why we have put you on our sustainable-growth panel."

Mitch had a sinking feeling. Later, after they settled into their lodging space, he expressed his fears to Jake and Janey. "They just don't get it. I can see their grand plans for exponential industrial development will carry such momentum that we will end up right back where we were in the early 2000s, after a century and a half of burying our heads in the sand about global warming effects from massive development in the *first* industrial revolution."

Jake responded, "And this is your chance! Your panel can create sound guidelines for growth with restraints. We are facing all of the bad habits from our emerging industrial revolution of the twenty-second century. We can nip these habits at its bud."

Janey added, "You can do what you can with the benefit of your hindsight into the mistakes of the past. You will also have the benefit of your new, enlightened, open-minded outlook that has been growing with every journey. We both trust that you will not turn these pioneers off with the arrogant and abrasive attitude that may have plagued you before. Remember, you may be the sole voice of restraint in a world that is hungry for rapid expansion. Unlike the Industrial Revolution, the science and technology knowledge of the twenty-first century is already with these survivors, so your challenge will be great."

The next day, Mitch was introduced to the fellow professionals who would serve on his panel. Calvin Curtiss was a civil engineer who served in the US Army Corps of Engineers. He was middle aged, short, and squat. He sported a crew cut and was dressed in what appeared to be a military uniform. He had a very serious—almost hostile—disposition.

Harley, a chemical engineer, was a bald man who wore heavy black horn-rimmed spectacles. He appeared professorial and Mitch hoped that this was a good connection point.

Hanaford was an agronomist, bringing an agriculture perspective to the panel. He was a tall man with black hair and hazel eyes, who seemed very quiet and hard to read.

Shelski was a Polish hydrogeologist. Polish by descent from a long time ago, but curiously retaining an accent. Mitch took an immediate liking to him and felt that with his own background, Shelski would have a kindred spirit on the panel. He was tall and lanky, he sported a handlebar moustache, he had an albino shade of blond hair and he had striking red eyes.

Bertrand was the most interesting of the group. He specialized in transportation engineering. He was a short man with an athletic build and a goatee. He wore a blue and gold beret. He had a wide knowledge of information outside of his field of transportation design. He spoke privately with Mitch after the introductions, providing him with more details of the past forty-five years as well as the evolution of plans for the exponential growth that had been spearheaded by Steel Town. Mitch asked Bertrand about the status of the rest of the world outside North America.

Bertrand shrugged. "We have no direct contact, though we presume that there is some semblance of society. According to one survivor of an enclave in Virginia, which collapsed in the 2070s, there was a fleet of three ships that arrived from Europe in 2060 that was able to contact the remains of the U.S. government just before its collapse. Allegedly, they told the Virginia people that Europe had adopted a network of their own enclaves that were able to remain in contact by a sophisticated network of radio communications provided by Britain and France. The European contingent supposedly informed the Virginia enclave that most of the nations of Europe had collapsed, with the exception of France, Britain, Poland, and Germany. However, the seeded enclaves were the main support of society. The European enclaves had focused on explorations to other parts of the world. Their main point of contact was with the kibbutz network of Israel. They had discovered that the tropics of Africa and south Asia were thriving on sustainable agriculture. These eco-communities were called *himaya agricural,* which is Swahili for agricultural empires. They established trade routes to Europe and Asia and bartered food products for technology, manufacturing supplies, and knowledge. The food supported Europe's enclaves and those of the remaining nations, which prevented the total collapse of their civilization."

Bertrand paused for a moment, then continued. "This history, however, was from a single source. It is not corroborated by anyone else. In fact, it was a story passed along to us from that time and to succeeding generations. So we have no way of verifying whether there was contact with a European exploratory fleet. There is a plan to build a fleet of ships to travel to England in a few years to seek out what civilization remains in Europe and elsewhere."

Outside of his knowledge of the fate of the outside world, Bertrand had an avid interest in science, and sensitivity to avoiding the same path of industrialization that had led the world toward a different type of climate disaster in the early 2020s.

"Mitch, we were spared from the disaster of an overheated world back then by the ironic sudden arrival of the climate-devastating eruption that gave us a frozen wasteland.

I fear that we are now on the threshold of re-launching a second industrial revolution that will lead us down the same track of global warming emergencies that we faced in the early 2000s."

Mitch responded, "You and I will be allies on this panel. We will need to act as wedges to the bulldozer of the world's intent to go full speed ahead." The two shook hands and parted.

Mitch had a different type of conversation with Dr. Shelski. He was also thin and lanky. It seemed that obesity in this food-scarce world was close to non-existent. Although he was trained as a hydrogeologist, he had also attained skills in climatology. "These idiots are planning on plowing straight ahead with the massive industrialization plan no matter what we recommend," he said. "What they are ignoring is the fact that CO_2 levels in our atmosphere are still hovering in the 450-parts-per-million range, somewhat higher than before the eruption. I actually expected that the levels would have been higher since the 150 years when we ceased our emissions, based upon analysis of scientists from the early twenty-first century, who predicted that it would take centuries for CO_2 to stabilize after emissions stopped.[50] Other studies claimed that the 'hangover' would last millennia[51] I guess that it must be that the soils, biota, and oceans have absorbed more of the CO_2 than predicted following the absence of human emissions during the period after the volcanic blast. This sucks the CO_2 out of the atmosphere."

Mitch's eyes opened wide. "Holy crap! I just realized. Since the volcanic sulphate veil has lifted from the planet, we are poised for another eco-disaster!"

Shelski gave him a knowing look. "It is no longer 'when' the veil rises. The aerosols are virtually gone and the planet is now just as transparent to the incoming sun as it was before the eruption. The CO_2 is still there and we are posed to belch out more, just like before the 2020 eruption."

Shelski and Mitch talked about the perspectives of the other members of the panel. Mitch told him about his conversation with Bertrand.

"That is good news! The three of us will make a near-majority block on the panel. But it will be no cakewalk in our efforts to get our views on restraining growth into the discussions. Almost everyone else on our panel will fight us. All of the enclaves want to surge forward with the exponential industrial development scheme come hell or high water. They want no restraints."

Mitch nodded. "You're right. We will have some internal resistance from our own panel. I spoke to Harley and he is deaf to any suggestions that we slow the intended growth momentum to give us pause for incorporating sustainability. I did not speak to Curtiss, but Harley hinted that Curtiss is with him."

---

50 See, for example: http://www.rachel.org/lib/zero_emissions_required.080227.pdf.
51 http://www.nature.com/climate/2008/0812/full/climate.2008.122.html.

Shelski confirmed Mitch's hunch. "Oh, Curtiss is even more recalcitrant. Full speed ahead is branded all over him."

Mitch replied, "That leaves Hanaford, the agronomist, as the wild card. My friend, we have our work cut out for us."

That evening, Mitch shared notes with Ter and Prelee, telling them what he'd learned from his conversations with the other panel members and expressing his concern about the unrestrained growth theme of this summit, as the world prepared to re-launch industrialization. Both Ter and Prelee were noticeably silent to this concern.

Then Prelee said, "Careful there, Mitch. This summit is aimed at getting society off the ground. We do not want to shackle our efforts with environmental restrictions."

The next day, the work of the sustainable growth panel began. Mitch was taken into a conference room promptly at nine. He greeted his fellow panelists, and with no pleasantries, Curtiss got to business, opening the discussion, immediately taking the lead role. "I will chair this discussion."

Oh brother, this guy has brass balls. I can see he will be a formidable opponent. Where is Reverend Tyndall when I need him?

Curtiss continued. "I have standing orders from President Prelee that we are to complete our business today and provide whatever recommendations that we might have to him so that he can incorporate them into the plenary session program on Friday."

Mitch objected. "That is a ridiculously impossible charge. I can see presenting our preliminary observations, but developing a science-based guidance system for the emerging industrial state in one day is impossible."

Curtiss responded, "Well, Mitch, that is the way it is. So I suggest that we move on. We will make our final recommendations by the end of the day. It will be a unanimous decision."   Before Mitch could respond, Curtiss gave his next order. "We start with putting our initial views on what we should recommend on the table."

Mitch shook his head and bit down his anger. *Hold on. I will hold my guns until the right moment.* He wondered what happened to Jake and Janey. They seemed to have vanished into the walls of silence during this summit. He had seen them only briefly last night.

Hanaford opened, "I can see how agriculture can be reignited, now that full sunlight has returned to us. The ash is enriched with nitrates and other essential minerals and nutrients. The ash soils have, over the past century, been colonized by lichens, mosses, and other simple vegetation that have restored carbon. The areas in the 500-mile radius around Yellowstone are still wastelands, being covered by ash so deep that organics from the buried soil would not be available to the sunlight for plant growth. I would propose that we start our agriculture recovery in the southeast and slowly spread it westward and northward. We need to immediately initiate the industrialization of agriculture with big farms and mass production of fertilizers."

Shelski objected. "Before we start forging ahead with big-farm agriculture, we need to step back and consider the alternative of sustainable small farming, at least as a complementary practice. Big Ag is one of the practices of the past that got us into trouble with mass production processes that depleted soils and released nitrous oxide as a heat-trapping gas. We have an opportunity to avoid the mistakes of the past and we need to—"

"You are out of order, Shelski!" Curtiss cried. "Our charge is to catalyze the reemergence of society through science and technology, not to impede it!"

Mitch wanted to belt this military man, but he decided to hold his main guns and only expressed his objection. "I want it noted that I object to having our self-appointed chair cut off discussion in an obvious attempt to railroad us into a 'roaring 1940s' approach to the re-industrialization of society."

He was backed by Bertrand. "I second Mitch's sentiment. By God, we will fall back to…"

"So noted," said Curtiss. "We move on."

"Wait a minute!" shouted Shelski. "I want to hear more of Bertrand's viewpoint."

Before Curtiss could cut him off again, Mitch took a deep breath and continued. "Gentlemen, we have a serious charge in re-starting civilization. We do not want to tumble into the same mistakes of history. I will defer to Dr. Bertrand to brief us on the state of our atmosphere, but suffice it to say that we are right back at 2020. We still have well over 400 ppm of CO_2 in our atmosphere. We should expect an about-face in our climate, now that the volcanic veil—that reflected incident sun heat compensating for the greenhouse effect—has lifted. The aerosols have only masked the greenhouse effect over the past 150 years, and now it will return."

Hanaford responded, "Aren't we supposed to be focusing on guiding us into the recovery of civilization? All you keep focusing on is slowing down. That is counter to our mission."

Shelski chimed in. "I never saw our mission as helping the world charge full speed ahead. The written orders for our panel were to provide scientific and technical guidance on *how* we re-emerge as a society."

Curtiss became brash. "But I gave instructions to catalyze our re-emergence as a society! I don't see the threat here. Earth's population is now only a small proportion of the nine billion people who existed before the volcano. How can we be a force for causing global warming to occur?"

Shelski put his argument on the table. "Who crowned you king, Curtiss? And, as a matter of fact, no matter what we do, global warming is upon us still. The CO_2 that we pumped into the atmosphere during the first industrial revolution is still here. With the volcanic veil lifting, it is only a matter of time before our planet starts overheating again! And how do you know how many people are out there? We have no contact with the outside world! The levels of CO_2 in the atmosphere are higher than they were prior to the eruption due to the first industrial revolution and the **CO_2** didn't wash out with the sulphate and ash veil. We now see our temperature rising 0.2 degrees Celsius since the veil lifted from the stratosphere. Even if we add nothing to the atmosphere, we will be facing the same warming crisis that faced the world prior to 2020. We can only hope that CO_2 levels will soon return to pre industrial levels of the nineteenth century sooner rather than later and in the meantime we will have to adapt. But if we start emitting CO_2 again, it will be like throwing gasoline on the fire."

As the heated argument continued with the panelists raising their voices and talking over each other, Mitch thought about Reverend Tyndall. What would that man have done to lower the temperature in this very room? He took a deep breath. After pounding his fist on the table to get everyone's attention, he used a calm voice to put his proposal on the table. "Gentlemen, gentlemen. It is understandable that our passions are high in this first meeting and we are getting stuck on a fundamental point of our mission. I recommend that we take a break from this session to think about our positions. We should re-convene in an hour to share—with open minds—how each of us would like to see society re-emerge. When we come to a consensus on this, we will move on to write up and discuss specific action steps to achieve those goals."

Curtiss broke in. "Sorry, Mitch, we don't have time for that. Now let's dispose of this nonsense and do what I originally said we would do—each of us putting our thoughts out right now on how to jump-start industry and comm—"

"I move that we act on Mitch's recommendation," Harley contributed. "We take a break for an hour and return ready to discuss our written goals."

Hanaford seconded the motion. The motion was adopted five-to-one in favor.

When they returned to the table one hour later, the heat had calmed down and Curtiss appeared to have shrunk in stature, as Mitch led the discussion, acting effectively as a moderator rather than an iron-fisted dictator. For the first time, Mitch noticed Jake and Janey in the back of the room. They both gave him a thumbs-up. Yes, there were significant disagreements—Curtiss was joined by Harley in promoting the acceleration of re-industrializing society, but Shelski, Bertrand, and Mitch were advocating the integration of massive scientific climate monitoring and gradually phasing in emission controls while industrial development occurred. They also proposed a one-year hiatus to develop a comprehensive plan to balance industrialization with sustainable growth guidelines. On the other side of the fence, Curtiss and Harley pushed for the full development of all industrial growth capacities with minimum restrictions. Hanaford—the swing vote—reluctantly took the full-growth position. With Mitch at the helm, they agreed to a comprehensive recommendation that compromised the Harley/Curtiss/Hanaford camp's goals with the position of Shelski, Bertrand, and Mitch. Despite this compromise resolution, Curtiss stubbornly clung to his uninhibited growth proposal and made a motion to recommend a full-speed-ahead development of business and industry with a re-evaluation of impacts in ten years. Harley seconded the motion. It was defeated by a 4–2 vote.

In the afternoon, they hammered out specific re-industrialization guidelines. Mitch orchestrated a balanced discussion that was interrupted by periodic outbreaks from Curtiss. The other five in the group were increasingly ignoring him. The final re-industrialization guidelines emerged in the mid-afternoon. They were stuck on Mitch's insistence (backed by Shelski and Bertrand) that development be tempered by green technology that would reduce CO_2 emissions by 80 percent per capita (once an accurate estimate of the world population was made) relative to 1990 levels of industrialization. This was countered by Harley and Hanaford's idea (backed by Curtiss) to allow industrial growth to occur unabated. Hanaford, however, thought that society should industrialize while, at the same time, phase in emissions controls as needed.

The group was in a 3–3 deadlock on this sticking point when Mitch called for another time-out to cool down and take stock of their stalemate. When they returned to the table half an hour later, they discussed and compromised. They argued their points of view, but listened to others without rancor. Mitch exerted a calming influence that moved the negotiations of viewpoints along smoothly. The other four on the panel ignored Curtiss's abrasive objections. He was clearly marginalized. By early evening, they had a "near consensus" for a compromise package of recommendations for the plenary session of the summit. It was "near consensus" because of Curtiss, who was entrenched in his opposition to the entire deliberation process and would not budge on uninhibited industrialization. The entire day, Mitch kept his own views muted, with the exception of the voting on specific points to the compromise package. He devoted his energy to mediating the discussions. Now, at this stage of deadlock, he expressed his views. He held strong to his insistence that the "full speed ahead" industrialization of society should be tempered with low per capita emissions of greenhouse gases. His preference was to require technology that would reduce emissions by 80 percent per capita relative to what they were in 1990.

In the end, the group compromised Mitch's 80 percent reduction standard down to 50 percent, along with rigorous inspection and enforcement guidelines. The rest of the package concerned technology recommendations that would empower agriculture, industry, transportation, and controlled commercial/industrial development. This was framed as a new vision of development that would emerge across the landscape of the planet. Despite Curtiss continued harsh objections, the compromise work of Hanaford, Harley, Shelski, and Bertrand (they were now dubbed by Curtiss "the gang of four") moved forward.

Not bad. I have to admit that the modifications that were made to my original proposal have made the package more realistic. In fact, it is now a stronger package.

When the question was moved by Shelski (and seconded by Hanaford), Curtiss insisted that they needed a unanimous agreement in order to present this to the full summit. The gang of four started shouting him down. Mitch pounded his fist on the table once. That is all it took to get the attention of the full panel. "We will take a break to cool off and return to the table in half an hour."

When they came back, Mitch presented several further compromise points that the gang of four had crafted during the break. Still, Curtiss stubbornly clung to his opposition to any controls on unfettered industrial development. At this point, Bertrand spoke. "Look, we have a motion on the table that has been seconded. I call the question to a vote."

Curtiss stood up and launched his final torpedo. "I oppose this proposal and the full process that led up to it. I am walking out. Our guidelines from Prelee called for a full vote from the six of us. See how far you get with this without my vote. You will be five." Then he left.

Shelski looked perplexed. "So what do we do now?"

Mitch, who had been vigorously taking notes, looked up. "Never mind Curtiss. His threat is empty. We ignore that bag of wind. The question was moved, seconded, and Bertrand called for a vote. Before he left, Curtiss registered his opposition to the motion, so I will record his vote as a 'no.' By the way, I consulted with Prelee, who advised me that our recommendation only needs to be by simple majority. So, now we vote on the motion." He repeated the motion that had been put on the table. All present voted yes and the vote was recorded as 5–1 in favor. The panel, minus Curtiss, then spent the better part of the night shaping their proposal into a full report for the plenary session.

The next day, they presented their report to the full Summit. The plenary board accepted the science and technology panel's report and passed their recommendation, but reduced the 50 percent cuts to 20 percent, (relative to 1990 per capita emissions) with the stipulation that society would further decrease per capita emissions by 5 percent per decade until the 50 percent target was achieved. The plenary board also rejected Curtiss's minority report.

Bertrand whispered to Mitch, "Man, Curtiss has gall in producing a minority report, but he is now silenced, once and for all."

After the plenary session, Mitch turned to Jake and Janey. "Not good—20 percent is not enough." After absorbing a long stare from the twins, he took a deep breath. "But it is a starting point."

Janey smiled. "You've come a long way, Dr. Mitch."

Jake added, "But now it is time to move on."

"I know," Mitch responded. The three of them jumped into a hologram bubble, lightly tinted with crimson, which appeared before them.

As their bubble floated away toward the beginning of the twenty-third century, Jake asked Mitch if he had any questions. He thought for a moment. "Yes. It is one of those 'I'll explain later' points that you both spoke of during this last stop. All of these folks in our travels seem to have been expecting us and know what I need to observe or where to inject me into the action. Yet they seem to know not to ask questions about my 'dimension.' What gives? What do they know about our travels and how do they know that we are from a different place or time? And how is it that President Prelee knows a bit more about my time tripping?"

Jake responded, "In moving from door to door and time to time, there is a phenomenon called 'universal consciousness.' It is what we refer to as 'the stream.' It exists even in your 2020 world. It is available to everyone, but only a select few tap into it. You have that sensitivity, Dr. Mitch. That is one reason we chose you for these journeys."

Janey took the baton from her brother. "When we visit these worlds, we purposely steer you toward those 'enlightened few' who are in positions of authority and who have this vague awareness that there are alternative universes that exist out there and that some people have the ability to skip from door to door and time to time in these alternate dimensions of space and time. That includes us, Dr. Mitch. We have the ability to select the gifted ones for you to visit in these dimensions. Prelee is one of those who is more sensitive to what is going on than most people."

Mitch remained silent for a while, absorbing all of this. Finally, he looked at his two guides. "What about your own sensitivities? Most of the time, you both blend into the background—seemingly aware of what we are facing. Yet there are times, such as the Yellowstone eruptions, when you seemed genuinely anxious. You seemed to revert from guide to in-scene participant of these realities. What gives?"

Janey responded. "As I explained, from those among us gifted with the insights of the universal consciousness, there is variability in the degree of sensitivity. Your old friend Casey, for example, was more sensitive than we are. Jake and I have, in our understanding, some—but not all—of the details that you are facing. So, in the case of the volcano eruption, we were not quite sure how things would turn out. We did not know for sure that any of us would survive the eruption and so that moment was as much of an adventure to us as it was for you. We do, however, know that we are destined to take you to 2224—your next stop."

Mitch again took this in with silent contemplation. "But now I am wondering about your mission—and mine—in taking me on these journeys? What am I supposed to do with all of this knowledge?"

Jake responded, "That is a story for another day."

Janey continued, "And now we must land in our next time destination of Door #6."

2224

The bubble that they were traveling in began to evolve into clear transparency. It occurred to Mitch that the crimson tinge to this part of the journey was fading, bit by bit, every step in time that they moved from era to era. Their now fully transparent bubble hovered over a large room. They were looking at a meeting of seven people in a conference room. The person at the head of the table was clearly the person in power here. He was dressed in a red robe that flowed to the floor. He also had red eyes. His demeanor was full of confidence. A nameplate at his place at the table gave his title: Erik II, CEO of the North American Empire. Next to him was a woman who was a younger version of Leah-Ann, his descendant. Mitch realized the date. *That is not Leah-Ann, it must be her granddaughter.* Her nameplate identified her as Angelica Warner-Stuart, Science Advisor. Next to Angelica was a younger man; he did not have a nameplate. He looked vaguely familiar to Mitch. All of the others around had titles as well. Dr. Che-Ling was the Steward of the Economies. Tetra was the Minister of Manufacturing, Trade, and Transportation. Hanaford was the Barrister of External Affairs. It was clear to Mitch that this group was akin to the President's Cabinet in his dimension of 2020.

Erik spoke, turning to Angelica. "So, Angelica, you have been my Science Advisor for four years now. Your talented cousin has been at your side all along. You and Drew have had your finger on the pulse of the climate of our planet. Brief me. What is the outlook for future threats?"

Angelica cleared her throat. "I will summarize using the opening statement of our recently issued report. 'The slow period of warm-up of the earth is over. We are now facing a tidal wave of global warming that will wash over us as unexpectedly as the volcano did in 2020.' It seems to me that with the exception of us in the science community, most of the world is sticking their heads in the sand, just as in the early twenty-first century."

Che-Ling, speaking for the economies of the world, rebuked Angelica, "This warming that you predict sounds like a repeat of what we heard in the early twenty-first century and look what happened. How can we take a new set of "the sky is falling" alarms seriously?"

Angelica responded, "You have to understand that science can only give us a snapshot of the future, based on the best available evidence, but that is all we have. I can tell you with every bit of confidence that the best available evidence gives every probability that the warming of our world will bowl us over soon, if left unchecked. No one could have predicted the eruption of Yellowstone in 2020. If it had not occurred, the 2020 years could have given us a climate disaster of warming instead of cooling."

Mitch, above, said to Jake and Janey, "She is talking too much in the science jargon of shades of grays. These folks need to hear it in a more 'black-and-white' frame."

Tertra, the Manufacturing czar, cleared his voice and spoke. "But Angelica, we are a planet that may have 1 billion people, less than 12 percent of the 2020 world. How can you compare now with then?"

The Emperor interrupted. "What *does* your report say about the population factor as it pertains to your prediction?"

Angelica responded, "I will turn this question to my deputy advisor, Drew, who did the analysis of our population and greenhouse gas loadings."

Mitch thought about the generation time lapse and whispered to Jake and Janey. "I get it, Drew is Leah-Ann's grandson. My descendant."

Drew was a handsome man with wavy black hair. He wore red glasses, and he had the build of an athlete. He exuded confidence. "You are correct that our lower population levels will give us lower rates of CO_2 emissions to the atmosphere in the near future. However, you are missing the 'bathtub' analogy. Our emissions are the faucet and currently are miniscule compared to pre-eruption years. But the tub is still almost full—the atmosphere is overloaded with CO_2 from the emissions of the post Industrial Revolution, which still linger in the atmosphere. Just a small increase may cause us to overflow. If you look at the rate of increase of our population, it will be only a matter of time before we catch up to the fast rate of growth in the nineteenth and twentieth century era. The most important thing for you to consider is the baseline. It took nearly 200 years to increase CO_2 from 275 part per million in the early 1800s to 410 parts per million in 2020. The volcano did eliminate the temperature increases, but CO_2 stayed in the atmosphere. Some washed out, but only a small amount. In fact, CO_2 was *added* by the eruption. In any event, the dangerous concentration of 440 parts per million of CO_2 is still present in the atmosphere. A small bump from re-industrializing our society without emission controls may push us over the tipping point into an eco-disaster. My analysis found that increasing our emissions from the emergence of our second industrial revolution will be significant. Most scientists agree that 450 parts per million is the 'tipping point' where the climate system will take on a life of its own, spiraling us into a world of severe climate impacts from a warmer atmosphere. As I said, we are now at 440 parts per million CO_2. The consequences will be similar to those of the volcano, but going in the direction of warming instead of cooling of the climate. It is already coming upon us. Just look at the loss of coastlines due to sea level rise in the last fifty years."

"Go get 'em, Drew!" shouted Mitch from above. "That's my great-times-eight grandson!"

"Shhhh!" said Janey. "We don't want to interrupt them."

"But you told me in the other world that they cannot hear us in this bubble."

Jake replied, "Yes, but they may sense us if you shout. Remember that stream of consciousness talk that we had on our way here."

Herb Hanaford, the equivalent of the Secretary State of the old days of the U.S., chimed in, "Yes, we lost coastlines, Drew, including Florida. Yes, the world lost the southern part of the Indo-Pakistan Empire to the ocean. But we are a small population; we can adapt by moving inland."

Angelica said, "With all due respect, Barrister Hannaford, we are only seeing the tip of the iceberg. You may have noticed that in the last ten years, we have more heat waves, more intense hurricanes, more tropical diseases. All of those events amount to higher fatalities and property destruction. Agriculture is starting to suffer from more droughts. Shall I go on?"

Above the room, Mitch shook his fists in the air. "Yes, Angelica! You've got him on the ropes. All facts, no conjecture."

Tetra broke in. "What about the solution of geo-engineering? I read about the concept from the early twenty-first century where we would adjust the amount of sun coming in to the earth by sulfate injections into the stratosphere or launching spacecraft that eject silver nanoparticles that reflect the sun. Our manufacturing sector could easily ramp up to that task."

Angelica responded, "Oh, I don't think you want to go there Tetra. We do not know enough about the climate system to tamper with the world's thermostat. As we have seen in the last 150 years, climate can flip from one state to another with just a small nudge. When we try to be the ones doing the nudging, we will most likely cause more damage."

Che-Ling, the Steward of the Economies, came back into the discussion. "I have read your report and it all points to the proposal to raise the emission reductions from 20 percent to 80 percent per capita. How much is that going to cost?"

Angelica turned to Drew to answer the question. He responded. "Initially, it will require a capital investment of two to three trillion for research and engineering. Then the actual deployment cost will be approximately fifty trillion. Finally, worldwide operating costs may run around five trillion per year."

Che-Ling responded with a shout. "Fifty trillion! That's more than our gross empire product! Why do we have to reduce emissions that much?"

Drew responded, "Back in 2064 this very issue was deliberated by a science and technology panel for the Great Industrial Emergence Summit. The panel recommended integrating emissions reductions with emerging industrial development. My ancestor Dr. Mitch Warner pushed for 80 percent cuts in emissions per capita, but the science panel recommended a 50 percent reduction package. Then the full summit reduced the commitment to 20 percent per capita. The higher 80 percent reduction of CO_2 emissions per capita was based upon science. That target represents the best available technology to moderate CO_2 emissions to levels that would have minimal impact on climate and avoid moving us over 450 parts per million CO_2. The 50 percent per capita reduction recommendation adopted was merely a compromise. The 20 percent cut is what we got—and as you see, it was clearly not enough. The panel's package also included a stipulation that this target be periodically evaluated to determine if further controls were needed. It has never been re-evaluated. This is the right time to crank the reductions up to 80 percent. As it turns out, we have only reduced per capita emissions by 13 percent since 2160 and more of the coastlines are being swallowed up by the oceans along with other environmental impacts."

Mitch beamed from above, clearly pleased with Drew's statement. "Atta-way, great-times-eight grandson. I only wish that you did not mention the 50 percent reduction number." It was as if Erik II had heard Mitch, though it is more likely that the Emperor picked up on Drew's mention of 50 percent, just as Mitch had feared.

"That sounds like the compromise we need at this time," Erik said. "We have to do something, but we cannot cripple our fragile economy. It is decided. You have a brilliant compromise, Drew. We increase the emission reduction of all sectors of our economy to 50 percent of 1990 emissions per capita. Tetra, instruct your science and technology team to develop the technology required to achieve the reduction target in our Empire. Che-Ling, you create the implementation guidelines that will phase in the technology at a minimum impact to our economy. Hanaford, you need to reach out to other empires around the world to enact a treaty to achieve worldwide agreement to internationalize the 50 percent reduction measures. Angelica, you create a package of climate-monitoring measures that includes a re-evaluation of the 50 percent reduction target ten years from now."

Surprisingly, Drew boldly interrupted this powerful emperor. "Before we move on, I suggest that we consider the fact that our current 440 parts per million is, by itself, a threat to society, no matter what we do with emissions. In Section VIII of our report, we recommend the deployment of Ambient Air Removal Technology (AART[52])" to effectively reduce the pool of CO_2 that is already there."

Everyone around the table, except Angelica, looked confused. Everyone was shocked that Drew was trying to add a postscript to Emperor Erik's decision.

Drew continued and pulled up a hologram that exhibited a giant structure of numerous fans. He explained that the fans are designed to pull CO_2 out of the atmosphere—scrubbing it clean and putting it into a solution that can be recycled for industrial use. "The AARTs are already engineered. In fact, this is a type of geo-engineering, but one that is designed to counter emissions and gradually remove CO_2 to levels that reflect what a natural world would be like without greenhouse gases. So, as we reduce our per-capita

---

52 Seehttps://www.technologyreview.com/s/531346/can-sucking-co2-out-of-the-atmosphere-really-work/.

emissions, the AARTs will work on the pool of CO_2 in the atmosphere and take us out of the danger zone. We recommend that society does the final engineering to determine how many of these units that we would have to deploy worldwide and then start building them."

Surprisingly, Emperor Erik did not explode at this challenge to his decision. He was silent—as was everyone else around the table—and thoughtful. Then he responded "Drew, I should throw you out of here with your brash, discourteous, and inappropriate challenge to my authority." Erik turned silent again and then continued. "But I have decided that your brashness is really courage. And it makes sense. Tetra, get the final engineering started along with a plan to deploy in five years. I want the construction started by March 1, 2229."

"Wow!" Mitch said from above. "Now that is one powerful emperor. It seems that they have dispensed with the legislative sector of government. It must be somewhat dictatorial, but they sure can get things done. Does this guy control the puppet strings of the other empires of the world the way he dictates the North American Empire?"

Jake and Janey smiled. Mitch took that as a "yes." She turned to the subject of reductions. "I know that you are probably thinking that 50 percent is not enough, but reconsider that position in light of everything that you have learned during tonight's journeys."

"Actually, I already have," said Mitch. "I realize that 50 percent is a leap and that change comes slowly. Also, I cannot totally discount the possibility that the impact of 1 billion people may not be enough to alter the climate. I think that the people from this dimension will indeed face future consequences by not going to 80 percent reductions, but I could be wrong. Unlike the twenty-first century, these folks have already achieved 13 percent reductions and I have no doubt that they will eventually achieve 50 percent. I would be more upset if it weren't for my descendant, whose bold and brilliant last minute assertion to deploy the AARTs was brilliant. I am so proud of Drew!"

Jake and Janey were beaming. Janey spoke for both of them. "Dr. Mitch, you have exceeded our expectations for your experience during tonight's travels to the three different worlds. You have become infinitely more enlightened in your ability to transcend conflict, compromise, and mediate, while sticking to your guns. We are proud of you. Now it is time to move on . . . I should say back . . . to your world of 2020."

Jake asked Mitch, "So, you had a question about the purpose of your journeys before we landed in 2224. Can you answer it?"

Mitch thought about this for a while. After about five minutes, he spoke. "You know, I do see the purpose of my being here. Every one of us holds some small part of the truth of the workings of the universe. When we hitch our passions to the universal consciousness flowing through each of our worlds, we see things in a different light. When we open ourselves to the perspective of others, we see a larger portion of what we might call the universal truth of all of our worlds. We grow. And we become more effective as agents of change. And, as for my specific mission, who knows? I may influence the broad stream of consciousness in the 2020 world. Perhaps my new consciousness will affect other worlds beyond my own."

Jake and Janey applauded in unison. She looked wistful, as if about to say good-bye. "And now, Dr. Mitch, we must return you to your home of 2020. As you drift back, think about what you have gained on tonight's journey. You have only one more door to visit."

I guess they are saying good-bye. Mitch felt sad that they would no longer be his guide for the last door. They faded away as the world around him drew into a warm dreamy state enveloped in a white puffy cloud that he floated on and that encased Mitch in an odor of lavender incense. He contemplated the entire experience of his journey of this evening. The cloud suddenly descended on a green pasture where there was a ring of many people holding hands. They were singing something in soft angelic voices. As Mitch slowly drifted toward them, the song sounded familiar and he eventually got close enough to this bright green landscape, ringed by crimson hills to hear some of the verses. They were singing "Across the Universe," a Beatles song from long ago.

Endnotes for Chapter 6

1. Given the errors of man's management of our planet and the consequences of countless extinctions throughout the ages, one must also consider the ramifications of nature. Be it through floods, droughts, plagues, meteors, sea-level rise, or volcanoes, nature is an important factor to consider as it continuously works through its ecological disturbances. However, just as nature has brought upon us mass extinctions, it also brings about the rebirth of new species and ecological balance. Indeed, many skeptics point to the fact that there have been ups and downs in temperature throughout history and the current warming is just one of those cycles.
2. Volcanoes come and volcanoes go. All of the volcanoes that we have experienced in our lifetime and most of the eruptions in human history are *not* super-volcanoes. There are six super-volcanoes that exist on our planet, including Yellowstone. However, one super-volcano—Toba—erupted 70,000 years ago and caused a 1,000-year cooling period. Human population was reduced to 10,000 people and this was considered a "bottleneck" in human evolution.[53] The magnitude of the 1992 Pinatuba eruption was miniscule compared to super-volcano eruptions of the past.[54]

---

53 See http://volcano.oregonstate.edu/toba
54 See http://volcanology.geol.ucsb.edu/gas.htm.

3. Although eruptions of super-volcanoes of the past are pieced together by paleoclimatological data, we can never know for sure what the effects may be. Therefore, the specific facts of the Yellowstone eruption here are part conjecture, but based upon plausible scientific prediction of the enshrouding of the earth with a deadly veil of ash and aerosols.[55]
4. Eventually, for all volcanoes, all ash falls out and the aerosols are washed out. How long it will take to return to normal depends upon how large the eruption would be. Slowly, there will be a return to conditions prior to the eruption. CO_2 in the atmosphere prior to the eruption may not be reduced significantly, depending upon the length of the cooling period, but some may wash out. Yet, volcanoes also emit CO_2 and the atmospheric concentration may increase after the eruption.[56]
5. The author's interpretation of the Beatles song "Across the Universe" is a means, through meditation, to connect to the universal stream of consciousness that exists all around us and that we have the ability to tap into. Carl Jung, the renowned psychologist, describes this theory as a collective unconscious, which collects and organizes personal experiences in a similar way with each member of a particular species.[57]
6. The concept of sucking CO_2 out of the atmosphere is not a fantasy. Dr. Peter Eisenberg, a physicist at Columbia University has constructed a working prototype that holds promise for the future.[58]

---

55 See https://en.wikipedia.org/wiki/Toba_catastrophe_theory.
56 See: http://volcanoes.usgs.gov/hazards/gas/index.php
57 See definition of "collective unconscious":
44https://en.wikipedia.org/wiki/Collective consciousness.
58 Seehttp://www.npr.org/2013/06/27/189522647/this-climate-fix-might-be-decades-ahead-of-its-time

INTERLUDE 2

Mitch felt a soothing sensation on his back. Someone was giving him a massage. From a world that seemed far away, he heard someone calling him. "Mitch! Mitch! Wake up. You are emerging from another night of dream journeys."

He opened his eyes. There was Leah, his wife. He was back at his home in 2020. Mitch blinked a few times. He had mixed feelings about being home. On the one hand, he felt that he was floating in from an enlightened journey that he didn't want to end; he already missed Jake and Janey. On the other hand, there was Leah. Not Leah-Ann, his descendant who was in Doors 3, 4, 5, and 6, but Leah, his lovely wife. There she was, smiling in front of him and holding a tray of breakfast. Breakfast in bed. Mitch spoke. "Well, as Dorothy said in *The Wizard of Oz*, 'There's no place like home.'"

Leah kissed him and said, "It's good to have you back, dear."

Mitch frowned. "I had one of those night-long dreams again, Leah."

"I know," she responded.

Mitch replied, "Was I tossing and turning all night, like the last time?"

Leah frowned. "No, it wasn't that. I think that I was with you on your journeys through three more doors."

Mitch's eyes opened wide. "What? Please explain."

"Okay," she said. "You had two people with you. Twins whose names were . . . " She thought a moment. "I think that the names were Jack and Jill."

Mitch laughed. "Jake and Janey."

Leah nodded. "Right, it was Jake and Janey. They were some kind of guides."

Mitch looked confused. "But how could you know that?"

Leah stroked her husband's curly hair. "Let's go slow on this, dear. It is coming back to me a bit at a time. Let it Be, for a moment."

It occurred to Mitch that it was Saturday. Thankfully, he did not have to go into work today and had the weekend to sort this all out. He ate his breakfast and had light conversation with Leah, but there was an elephant in the room. Finally, Mitch returned to the burning questions in him. Then he relaxed and shook his head.

"I know. I must have been talking profusely in my sleep all night and with a lot of clarity and detail."

Leah smiled. "No, dear. I was in your thoughts. For example, at the very end of your journey through the 'geo-engineering door' when you sensed that you were seeing Jake and Janey for the last time, your thought was: *I guess they really are saying good-bye.* You see, I was there as an observer, and obviously not there as a participant. I was conscious of your journey the whole time. Maybe I just intuited your whole experience through the 'stream of consciousness' that you kept talking about with the twins. All I know is that I picked up your vibes and saw through your eyes. I liked it."

Mitch's eyes widened again. "My God, Leah. You picked up the internal dialogue of my thoughts? You really were there with me! It's funny; I don't remember having any feeling that there was someone with me all night. Although my intuition tells me maybe I did have some sense of your presence with me. Maybe that's what made my transition to home much easier this time."

Leah nodded and smiled. "For me, too. Knowing and feeling your experience gives me so much relief. I must say, Mitch, your return from the first night of dreams scared the bejeezus out of me. You were so off-kilter last time." She looked at him. "There is something different about you now. I can't quite put my finger on it."

Mitch smiled back at her. "Enlightenment, my dear. When I returned from the first night's journey, my soul was on fire with passion. I felt that it was my mission in life to awake the world to the choice of fates that we face if we do nothing, something or everything to counter our changing climate. After last night's journey, my soul is still burning with passion, but with a controlled burn of understanding that there are other ways of looking at things. I am, in fact, looking at the universe with new eyes."

Mitch and Leah, who was curled next to him, were quiet for a long, tender period. No hurry. No confusion. It was just nice. However, there was still something unsettled within him. With his new eyes, he didn't know how to interact with the world. He knew that there was something that he needed to do; he just couldn't put his finger on it. He decided to let that thought be. He stroked Leah's jet-black hair. "So, what did you think of my—I should say *our*—journey?"

Leah hugged Mitch. "Quite an eye-opener, but on another level it was quite entertaining. I shared your affection for Jake and Janey. I wish I could have met Casey on your first night's travels. I felt scared for you—I should say *us*—trekking across Antarctica through Door #4. I felt sad with you when you floated away from the 2164 Thanksgiving meal through Door #5. I was so proud of you in your interactions with the science panels of Doors #4, 5 and 6. Oh—and I was charmed by the fact that your great- great- great- great- great- great-granddaughter shared my name. Oh sweet Leah-Ann—she was my descendant as well. I was proud that she was named after me. I wonder if that was coincidence."

Mitch kissed Leah on the cheek. "I doubt that she had your name by accident. They seemed to know a lot about their history regarding us."

They lay silent again as the yellow rays of the rising sun flooded through the windows. Mitch furrowed his eyebrows and frowned. "You know, I don't quite know what to do with the wealth of all of these insights in our present-day world. I guess that you must have the same burden." They lay in silence for a few moments.

Leah responded, "Well, Mitch, we both got this same advice throughout our journey together. We must Let it Be. We should not over-analyze this. We should let our insights be our guide."

They were interrupted by Susannah calling from downstairs. "Mom, remember that you are going to drive me to my soccer match this morning. We've got to leave in half an hour." So Mitch and Leah started going about their normal lives. That first day seemed to last forever, as they both carried the burdens and joys of bearing multiple worlds on their shoulders. But for both of them, it was a calm sail and they both knew that there was no rush in sorting this all out.

At the dinner table that evening, their threesome family had their customary Saturday night pizza party. However, unlike their typical pizza parties, Susannah did all of the talking. Mitch and Leah were both lost in thought. Finally, Susannah became quiet. She looked at Mitch, and then looked at Leah, shaking her beautiful long blond hair in bewilderment. "Okay, guys, you need to tell me what is going on. Are you both stoned?"

Leah and Mitch laughed and Mitch responded. "No, Suzie Q, we both had some very vivid dreams last night."

Leah decided to share a bit of their secret. "Suz, we had the odd experience of sharing the same dreams. You probably cannot understand that but it may be one of those experiences that we will never figure out. They were nice dreams, though. Nothing to worry about."

Susannah looked at her parents. "You mean that you dreamed the exact same dreams?"

Mitch answered, "Sort of, Suzy Q. I think that we are still figuring it out, but we are mostly letting it be."

Susannah took another slice of the pizza and looked at her parents again. "I thought that I told both of you to call me by my full name instead of those stupid nicknames. Susannah. It's a pretty name. But getting back to your dreams, I think that it is kind of special. It may mean that you two are so in love that you are thinking and dreaming as one. You're not going to run to a shrink about this, are you? You shouldn't over-analyze this special experience."

Mitch and Leah laughed. "No, Susannah," Mitch replied. "Nothing like that."

His daughter smiled, took a drink of cola, and said, "Cool." She was off on her next stream of thought, whatever that was. Mitch felt nothing but love toward his daughter.

She was so beautiful: tall, trim, graceful as a gazelle, and those beautiful blue eyes and long blond hair. She was sixteen going on thirty. So smart. Maybe a little on the wild side, but she would settle in as a very capable adult some day. Mitch held back tears from his eyes. *Whatever happens, I don't want to lose this wonderful family. I am blessed.* He looked at Leah and had this eerie sense that she was thinking the same thought.

Sunday was restful. A perfect way to decompress. More so since Mitch and Leah experienced the same night-long journey on Friday. Susannah kept telling them that they were still acting weird.

On Monday, Mitch went to his campus office early. Lonnie greeted him with a concerned look. "Are you okay, Mitch?"

He wondered if she'd picked up on something about his new way of looking at the world. "I'm okay," said Mitch.

Lonnie followed up. "You know, I am always here to talk to you if you need a friend to lean on."

Mitch responded, "Lonnie I appreciate all of your support. Everything came to a head on Friday and I know you're concerned about my meltdown. Since then, something has happened that has given me a new way of looking at the world. No time to talk about it now, but let's have lunch together, and I'll tell you all about it. What do you say?"

Lonnie beamed as she brushed back her blond hair. "It's a deal." Mitch then went into his office and prepared for the morning staff meeting.

The first faculty member to enter the room for the staff meeting after Mitch was Dave Cornelius, who gave him an obvious cold shoulder. Mitch came over to Cornelius and extended his hand. "Peace, brother. I want to apologize for my confrontation with you last week, Dave. I have been under a lot of strain and I took it out on you."

Dr. Cornelius looked both surprised and disarmed. He shook Mitch's hand. "Well, it takes two to tango. I know that I have been goading you on the issue that we should probably not speak of again, Mitch."

Mitch took a deep breath. "Well, we should at least agree to one ground rule should we ever want to discuss it. We are fellow faculty members and friends first and climate change antagonists second. That will allow us to break off the discussions before they become heated, if we start down that path again."

Lonnie looked at Mitch with wide eyes. *There is definitely something different about his outlook today.*

The meeting started with the Department Chair, Dr. Alvin Klertz, handing out two memos to each faculty member. Mitch looked at the first memo, which was titled "Faculty Guidelines for Applying for New Grants." The second memo was "Faculty Guidelines for Participation in Organizations Outside of the University." Mitch smiled. He knew that both memos were aimed at him.

At the other end of the table, Lonnie was ready to take meeting notes when she received the papers. When she saw the subject of the two memos, she groaned inwardly. *Brace yourself, girl. Klertz is poised to goad Mitch into a blow-out.*

Klertz opened the meeting by asking them to take a minute to review the first memo regarding grant applications. It directed each faculty member to prepare an application request to the department chair that would include a statement of how their proposed research would enhance the department's goals and a separate rationale that itemized the administrative costs of conducting their research if the grant was funded. Everyone around the table knew that the new guidelines were aimed at Mitch and his maverick habits.

Dr. Ruth-Ann Gere immediately tore into the guidelines. "This memo is an insult to the independent integrity of scientists who are trying to explore how the world works. You as head of our department should be encouraging, not holding down our efforts with these so-called guidelines!"

Gere was backed by Dr. Tom Troutbrook. "Alvin, we had these discussions with Dr. Hannigean before he left as chair. We had discussions over several weeks and hammered out reasonable guidelines that were then approved by the dean. They have worked well for us."

Klertz cut into Troutbrook. "So say you, Tom, but I am chairman now and you will follow my guidelines, not Hannigean's."

Dr. Franklin Dyer joined the melee. "I have been in this department for forty years and I must say this is an outrage!"

Dr. Harold Tussock chimed in. "What precipitated all of this, Dr. Klertz? I am equally outraged."

The heated argument went on and on. Mitch remained silent. Finally, he tapped his hand on the table and asked to speak. He looked around the table in amusement. Klertz was poised to pounce on him. The rest of the faculty seemed to be poised for an outbreak from their fellow faculty member.

Then he spoke. "We are getting nowhere with this rancor. Why don't we take a five-minute break and cool down. When we return to the table let's think with an open mind and we can resolve this."

They took the breather. When they returned to the table, it was Mitch who subtly mediated the discussion, with deference to Klertz. They each put their views on the table and they all agreed to review the guidelines over the next two weeks. They agreed to be prepared to discuss them again.

Klertz wanted the final say. "But there will be no grant proposals submitted from our department until these new guidelines are issued."

Seeing the heat in the room rise again—they all knew that the guidelines were aimed at Mitch's new grant application—Mitch cut off a new confrontation at the pass. "I have an idea, Dr. Klertz. Why don't you and I meet separately to discuss my latest climate change grant proposal and use these guidelines as sort of a pilot to see how it would go?" There was a momentary silence in the room. Mitch sensed a surprised look of "what the hell?" from each person. Dr. Troutbrook's mouth was actually agape.

There is something different about Mitch, Lonnie thought. He had, for weeks, refused Klertz's demand to meet with him about his grant proposal and his demands that Mitch make this project more on the theme of pure science research. *There's something very Zen about Mitch's new behavior, but I'm not sure I like it.*

The discussion on the second memo followed in a similar vein. Klertz's requirement to review each faculty member's affiliation and interactions with student and outside activist organizations was obviously aimed at Mitch's involvement with such groups over the past several weeks. But before any faculty member could react, Mitch beat them to the punch.

"With all due respect, Dr. Klertz, these guidelines are clearly aimed at me. I do not want the others in this room to be saddled with these new rules unnecessarily. Why don't you and I meet to discuss how my recent activities outside of the University may be modified in a manner that will not embarrass you or the department?"

Again, all were in awe of Mitch's non-confrontational behavior. Klertz was not only disarmed by this new attitude, he was bewildered. "Well . . . I guess we should do that. If you can convince me that these guidelines are not needed for you and other faculty members, well maybe we can dispense with this second memo."

After the meeting, Dr. Gere came up to Mitch. "We were all trying to back you and all you did was back off."

Dr. Franklin Dyer just shook his head and said, "Mitch, don't worry about Gere. She is just as confused as the rest of us. We don't know what to make of this new guy inside of you. Personally I kind of like the new mediator that has emerged, but it is just too different from the old Mitch. Hard to swallow."

Dr. Cornelius joined them. "I, for one, like this new Mitch."

Mitch and Lonnie went to lunch in the campus cafeteria. She opened the conversation bluntly. "What gives, Mitch? You looked like you were declawed in there."

Mitch smiled. "I can best explain my new perspective to you by telling you about my dream experience of the other night."

Lonnie replied, "Oh, I should have known. Your behavior has taken the same sudden turn as it did after your first set of 'journeys,' as you called them."

Mitch told Lonnie about the Friday night of dream journeys, including Leah's awareness of his adventures.

Lonnie expressed her concern regarding the "about-face" in Mitch's personality. "It seems as if I have known three Mitches over the past month. There is the Mitch before the first night of dreams, the 'wild-eyed' Mitch that came after the first set of dreams, and the declawed Mitch sitting in front of me now. I am concerned about you."

He looked at his secretary and friend. "I share your worries. I am wondering if my head will be able to handle these changes. I need to integrate everything to make myself whole again, but for now, I am just letting things be. I will be taking that vacation to the Utah canyon parks with Leah and Susannah during spring break, so that our family will have time to figure this all out."

Lonnie replied, "I hope that you can handle all of this in light of how you broke down after the first night of dreams."

Mitch looked at her and paused before he spoke. "Lonnie, what do you make of all this? Do you believe that the two nights were just nights of dreams?"

Lonnie nodded. "I am open-minded enough to think that your experiences were real, at some level, and they were way out of the ordinary. Particularly since Leah shared your experiences. But please—be careful and let me know if you or Leah start running into trouble."

He stroked his gray curly beard and said, "There's one more thing, Lonnie. I will be taken on one last journey through one last door."

She stared at Mitch. *Oh, no! Can I take another one of these mind-bending events? More importantly, can Mitch?*

Over the next several weeks, it appeared that Lonnie's intuition had been prophetic. Mitch became more and more ill at ease. He was not given to the same "meltdown" loop that he'd experienced after the first night of journeys, but he kept asking himself, "What is my mission, anyway?" Although he'd already answered that question during his return to 2020, he still could not figure out what his specific role should be in carrying forward the observations, experiences, knowledge, and enlightenment that he'd gained from his most recent journeys.

During their vacation in Utah, Leah shared with Mitch a similar feeling of discomfort. He didn't get it. Jake and Janey had done a good job of explaining how these journeys worked and had guided him to a comprehension of his general mission. But he did not know what to do with all of this in his world of 2020. Leah was more concerned about what would happen to her when Mitch took his trip through the final door. Susannah, on the other hand, was concerned about both of her parents. She sensed their unease and was disturbed about the distance that she felt from both of them. All in all, it was not a very fun vacation for any of them.

Lying on a couch a week afterward, Mitch thought about his disquietude. He couldn't fathom what he was going to see on the other side of Door #7 and he wondered when he would be taken there. He dozed off into a deep sleep.

CHAPTER 7

THE BLUE DOOR (# 7): THE GRASSROOTS SOLUTION

Can citizens working together at the grassroots level bypass government and solve the problems of climate change on their own? Is such a future too utopian to imagine?

***

2034

Mitch felt an odd peacefulness in the knowledge that there is hope for our planet, and well, for humanity itself. Having traveled so many different roads and had so many different experiences on this journey was mind blowing! With a head full of possibilities ranging from the absurd to the plausible, the hardest part for him has been to stay receptive to all these ideas. Upon further reflection, Mitch stopped. *Am I stubborn or what? Sometimes I wish I were back in the cabin where my family vacationed in the mountains of Ouray, Colorado in the 1970s. The air was so clean and crisp you could feel the essence of how it used to be. And the river, oh it was magnificent! Ah yes, I often wonder how I ended up in Miami! Well, no need for environmental careers in the remote mountains of Colorado! I hope to remain open to my next quest on my final journey.*

The new journey had begun. Mitch felt himself floating on a blue rowboat. The boat glided through a blue archway that enclosed a door that was suspended over the water. A sign on the blue door read Door #7: The Grassroots Solution.

The boat then drifted through the door into a cove. The weather was perfect. The warmth of the sun felt wonderful on a clear, dry day—not too cold, not too hot—and suggested late spring or early summer in Florida. The water of the ocean was turquoise. The boat was moored in the cove. Mitch had tied it to a tree. The waves rocked the boat and he felt very dreamy. There was no one around and there was no noise of civilization here. Mitch felt like sleeping. Suddenly, he bolted upright.

Wait a minute; I am in a dream—I'm sleeping already. Then he realized that this was the final door entering a new dimension—not a dream. Mitch smiled. *I shouldn't look a gift horse in the mouth. I just hope that whoever or whatever is controlling these journeys isn't casting me out here on my own.*

He looked out at the calm ocean. In the far distance he saw a spot on the horizon that seemed to be heading his way. A few minutes later, he could see a vessel floating toward him. Mitch noted that the boat was also the same blue color as his own boat. When the vessel approached he noted that it was not a rowboat, but a wider boat. A tall man dressed in a brown-hooded robe waved to him. Mitch waved back. He couldn't make out his face. The man—yes, it was a man—steered the pole-boat into the cove where Mitch was. Mitch noticed that the hull was stenciled with the numbers 2034. He still could not see the face that was enshrouded in the hood of the brown robe. Mitch helped the man onto shore and looked up in surprise.

"Reverend Tyndall!" He gave him a bear hug. "So, my guess is that you are going to be my guide for this final journey of mine."

The Rev dried his sandal-clad feet with a towel. "Well, most of the way."

Mitch recalled his childhood fun of skipping stones and reached down to find the perfect flat stone and sent it across the water. "Well, you are a sight for sore eyes. Tell me what is going on." Mitch looked at the number on his boat. "My guess is that we are in the year 2034."

The Rev smiled. "Right you are. Are you ready to get started?"

Mitch smiled. "Yes. Let's get the show on the road."

The Rev explained to Mitch that he was through a door of reality that no one back in his dimension would think as plausible. He told Mitch that he would be in the active participant role. We are going to Boston tomorrow. There you will be participating in a Council of Technical Advisors to discuss the mission of a national network of grassroots organizations who are taking the fight against climate change into their own hands. This will include local environmental sustainability groups from around the nation, sustainable farmers, enlightened industries, insurance groups, and others—even local government officials.

Mitch yawned. "So how did all of this get started?"

Tyndall replied, "Bored already?"

"Oh no! This is such a peaceful place, it makes me sleepy. Guess I need this."

"That is good, Mitch. This actually started in your dimension of 2020, with local land trusts, teachers, religious organizations, outdoor recreation groups, and sustainable living organizations, which were more and more skeptical of the prospect that the national government or the U.N. would ever act on climate change. The attitude was 'Well, if they won't do it at the top, we will do it from the bottom.' This also resonated in Europe, which already has their own homegrown movements. The common theme of each group was striving to reduce each individual carbon footprint. The hope was that this attitude would spread throughout society like a friendly virus. In the mid-2020s, there was momentum from organized coalitions that brought in sustainable farming that had remarkable success. They redesigned and built an urban transportation infrastructure that used electric cars, electric bicycles, dedicated bicycle paths, mass transit, and work-at-home initiatives. Some of the grassroots organizations created their own utilities that sold electricity to their customers and the grid exclusively from solar, hydroelectric, wind, and geothermal power. Then businesses, large and small, started joining in the movement. All along, local governments were allies with the coalitions and in the case of some of the larger municipalities, provided subsidies without any strings attached. The local grassroots groups would be in control. Insurance companies added their weight (recognizing that claims would continue to increase as the climate emergency unfolded), providing lower rates for customers in communities that made commitments to the common primary goal of all the community groups: the 80 percent reduction (relative to 1990 levels) in carbon emissions from each individual, each community, each business. There is so much more, as you will see."

Mitch shook away any sleepiness from his mind. "Now we're talking! The feds were never going to be able to do this. The U.N. efforts were a joke. With the exception of a few progressive states, not much was happening at that level either. Right down on the bottom of the ladder is the foundation of real change. As they said in the 1960s, 'power to the people.' That sounds corny, I know, but that is where big changes often start blossoming. How has our national government been taking all of this?"

Reverend Tyndall laughed. "Well, as you say, change—even political—starts at the bottom and rises up. The pressure from the grassroots level surged upward. With each succeeding election, more Congressmen sympathetic with the grassroots creed were elected. Then two years ago, an amazing thing happened. A third party—the Blue Sky Party—took control of the reins of power. President Rother was elected on the groundswell. She had a simple campaign dogma: 'Think globally, fight locally.' By this time the local green block had a majority in the Senate. However, the bickering and gridlock still continued on—particularly in the House. Of course, the worsening climate contributed to the groundswell. The economic woes from the collapse of U.S. farming, the wildfires, the droughts and the sea-rise inundation of our coasts all increased frustration with the lack of national governmental actions and the growing hope that the problem may be resolved with this grassroots movement."

Mitch looked very enthused as he rubbed his hands together. "I was wondering about the status of climate."

Reverend Tyndall pointed to a series of atolls in the distance. "Those small rock islands are what are left of the Florida Keys. It's much worse in other areas of the world, but you already know the consequences from some of the other doors you have gone through. In fact, at this summit we are going to, you will find very little talk about how bad things are.

One of the credos of the grassroots groups is to 'Let it Be on what has happened' and 'let it grow' on what we are re-creating.' Mitch was stunned by the possibility that society could be so positive. *It's about time.* "So, are the feds attending the summit?"

Reverend Tyndall took a swig of his lemon elixir and offered another bottle to Mitch, who accepted with a tip of his hat. He noticed that the bottle had an unusual texture—it was sort of a glass substance but lighter, with a crenellated surface. Reverend Tyndall noticed Mitch gazing at the glass and feeling its unusual texture. "That bottle is one example of how we are living these days. Buying containers of beverages from a supermarket is fading away. You will see that you will not want to throw it away. There are refill stations that have replaced soda machines. Many folks are making their own beverages—beer, elixirs, wine. So reduction, reuse, and recycling are taking on a whole new meaning. But let me get back to your question. The national government wanted to jump into the lead. Even President Rother—she offered to take control of the network of local community groups and subsidize the efforts to elevate their actions to a higher level. In a major initiative that she launched last year, she worked to build support by embracing and assisting the efforts of the local groups. Incredibly, the work of the international grassroots movement stabilized CO_2 emissions in ten years and they are now starting to decrease. So, President Rother presented a plan to support the grassroots movement in a new federal network with a goal to reduce emissions by 10 percent per decade."

Mitch shook his head. "Well, it sounds like a tempting offer, but I would be suspicious of the government offer if I were part of the network."

Reverend Tyndall patted him on the shoulder. "You've got that exactly right. Although there was no one national spokesperson representing the grassroots group, the response from individual groups was nearly unanimous across the nation: 'Thanks, but no thanks. We will gladly accept your subsidies, as donations, but we will retain local control. We will work every day at reducing emissions, but do not give us a ten-year plan' "

"President Rother came under increased pressure to ignore these responses and set up the government's own national network that would work with the existing grassroots networks. Her considered response was a surprise to everyone. 'We will not interfere,' Rother stated in a national address. 'In fact, I will introduce legislation to support the grassroots networks financially and stay involved only through federal coordinators and achievement standards, which will determine the eligibility of participating communities to future block grants.'"

The Rev took another sip of his elixir. "After months of negotiation, a newly formed council of grassroots organizations agreed to the concept, as long as the coordinators participated as advisors only. The president agreed to those terms with only one condition of her own. She insisted that the national council be elevated to a stronger and more active role, which would be supported by annual subsidies. This condition met with such resistance that it almost broke the deal. The deal was saved by a written commitment in the legislation that the national council of grassroots initiatives serves simply as a clearing-house and coordinating network with no intent of holding authority over the local groups. The deal was made and the funding is making its way through both houses of Congress, and is expected to pass."

Reverend Tyndall watched Mitch as he gazed out over the horizon. "So what do you think of all of this, Mitch?"

He kept gazing at the missing South Florida Keys, almost as if he did not hear. The Rev. "I think that before you started telling me all of this, I would have opted to stay here another week or so to soak up the rays, but now? I am ready to jump into the fray."

"Okay!" responded The Rev. "Let's go. Be prepared to jump into active mode. You will do very little in the observer role here, Mitch. You are part of the action. Now let's jump onto the blue cloud."

A blue cloud indeed floated toward them and they jumped on. Mitch was in such a mellow, thought-filled trance the he never knew when the surreal float on the blue cloud across the ocean ended and how he ended up in Boston. But here he was. The historic New England character of the city was always a pleasure for him to see.

"So, where is this summit, Rev?" The Rev looked at Mitch. "The Board Summit, which will be Friday, is in the Boston Civic Center Hall. You will be meeting at Faneuil Hall with five other specialists who represent a cross-section of environmental professions. Your group is being called 'The Grassroots Planners.' You will have four days to formulate a plan of integrating local community groups into a national coordinating network. At the summit on Friday, local grassroots representatives will be joined by representatives from green industries, local governments, and, of course, the EPA. They will base their meeting on the report that your group will submit to them that should have recommendations on how to launch the local grassroots initiatives into a national network. So, you six will have three days to produce a very important report."

Mitch and Tyndall continued to walk. "Rev, what is the objective of your summit?" Tyndall looked at Mitch and smiled. "*Our* summit—not my summit. Keep in mind, Mitch, that you are part of this, too, you are now in your active participant role. This is the 'coming-together' moment that has been planned for two years. The national network of local groups will be organized today. You will be establishing the mechanics of how the local groups—which we are now calling 'green buds'—will be able to enjoin resources and co-ordinate local actions into a national web of joint activities and shared technologies. We are also planning a 'coming out day'—the event for the people—next Earth Day."

Mitch and Reverend Tyndall entered Faneuil Hall and were greeted by two familiar faces: Bertrand and Shelski from the geoengineering panel of Door #5. Mitch lit up. "Hey, my trusted allies! We fought the battle to put geoengineering in its proper place; now we are facing a new challenge!"

Bertrand and Shelski looked at each other in confusion. Shelski spoke for the both of them. "Geoengineering panel? We do not know what you are referring to."

Tyndall whispered to Mitch, "Careful! We are in a different dimension and a different time from Door #5. Do not go there. This dimension is different from the 2034 dimension of Door #5."

Mitch looked embarrassed. "I keep forgetting that we are not in Kansas anymore." He looked at Shelski and Bertrand. "Forget it. I am thinking about a twosome I knew in another era that looked like you. So, I am pleased to meet you. Will you be briefing me on what we will be doing? Also, will you tell me about my role in this working group?"

Bertrand looked at Mitch. "Somehow, you do look familiar to me. Perhaps our paths did cross somewhere. Anyway, we are starting our first meeting this morning to discuss our goals in launching this national coalition and our guidelines in setting up a support network. We have been told that you and Reverend Tyndall have experience from a different time and place in mediating diverse perspectives. We sure will need that. You will be working with professionals that will represent a cross-section of the various sectors involved. By consensus, we will be asking you to chair these pre-summit planning sessions. We were told of your sterling reputation in moving groups with antagonistic viewpoints to a consensus. We were also told that you are from some other place and we know not to ask about your specific roles in that place—or time—wherever you came from. We intuitively knew that you should be our chair, so just jump into that role. Come; let's meet the other panelists. There will be six of us working together. One word of caution: an EPA representative, General Curtiss, will be joining us at the table. He is an arrogant, domineering fellow and don't be surprised if he tries to usurp your power."

Mitch turned to Tyndall. "Shouldn't you be moderating this meeting?"

The Rev answered Mitch's question by addressing the entire group. "If you are wondering why I am here, I am a mentor to Mitch. I am also available to all of you for advice if you should need it."

The Rev smiled and spoke privately to Mitch. "You have become exponentially enlightened in your journeys. Just get in and do your stuff."

Mitch joined the rest of the group at the table. He was offered an elixir beverage and poured a grape flavor for himself. The faces were familiar. It came to him as they approached Hanaford and Harley. He said nothing, as he knew that they would not recognize him from another dimension.

However, Harley surprised him. "You look awfully familiar. Have we met somewhere?"

Mitch smiled and shook his head. "Maybe in another time or another place." He looked around the table and counted. "Aren't we supposed to be six?"

Bertrand took a drink of his elixir. "It is Curtiss who is missing. He is the EPA coordinator. He will be here shortly."

Mitch groaned inwardly. *Won't Curtis want to dictate control of this group? That's like putting the fox in charge of the chicken coop.*

Shelski responded answering Mitch's thoughts, "Well, Mitch, we were persuaded by Reverend Tyndall that you have vast experience leading groups like this from whatever time and place you came from. That's good enough for us, though Curtiss clearly wanted to be in charge. So we are nominating you as our Chair."

No kidding. Mitch reflected that Reverend Tyndall could persuade anyone of anything. He hoped that he could meet this same bar of mediation. "Okay, I accept that charge and thank you for your vote of confidence." The door to the conference room opened.

"So, what is it that you are accepting?" Curtiss, an engineer from the US Army Corps of Engineers, approached the table and introduced himself. He looked at Mitch. "I have a funny feeling that I know you . . . from some other place, some other panel. I am sorry to tell you that my intuition tells me that I do not like you from whatever past experience that we shared."

Mitch raised his eyebrows and took a deep breath. "Well, whatever it was, how about we let bygones be bygones." He extended his hand. Curtiss grunted.

Bertrand looked at Curtiss. "We briefed Mitch about our decision to have him lead our proceedings and he has agreed to chair our group."

Curtiss glared at those around the table. "You know my feelings about this. It is the EPA who should have the lead role here."

Suddenly, a change appeared in Curtiss's demeanor. "Well you know, Mitch, I have been rude here. Chalk it up to a rough day. What do you say we start over?" The others around the table looked warily at each other but Mitch smiled at Curtiss and gave him the thumbs-up.

Mitch called the meeting to order. "Gentlemen, we can make this simple or complicated. I will recommend the simple way. This coalition can always fine-tune the process with more detail in time to come. I suggest that we devote today to composing a statement of cooperation between the local grassroots groups and the ground rules for the alliance. Tomorrow, we identify areas of cooperation. Wednesday, we can specify a process for ongoing communications and mutual aid. What do you all think?"

Bertrand, the transportation specialist, responded, "Mutual aid. I like the way that sounds. Why don't we title this document Mutual Aid among Grassroots Communities?"

Shelski added: "How about referring to the network of grassroots groups as the 'Down to Earth movement?'" They all liked that.

Harley, the civil engineer, gave his views. "We need to be careful not to be so general that the document is a paper tiger. We need to have some specificity to the nuts-and-bolts projects that we want to share and implement."

Hanaford, the agronomist, added, "I agree with Harley. The agriculture projects that we enact will be tricky as we transition to local technology."

Curtiss spoke. "And the EPA can play a major role in helping get this initiative off the ground."

Objecting abruptly, Shelski spoke up. "Wait a minute! The beauty of the grassroots movement is that it grew from the bottom with no bureaucratic structure holding our hand and keeping us down. I say that we just Let it Be and allow the ideas and technologies to emerge as they have been over the past decade."

Bertrand added his voice to the fray. "We do not need to reinvent this successful process. We, at the grassroots level, are accomplishing what government hasn't been able to do for years. How much have we reduced carbon emissions, Shelski?"

The Polish environmental scientist responded, "Close to 20 percent, according to recent monitoring."

Bertrand continued, "I rest my case. We do not want to re-create this into a new bureaucracy. Let it Be!"

Curtiss surprised everyone with his response. "Well, we seem to have a consensus on this point, Mr. Chairman. Okay, let's keep the EPA out of this for now. Let's keep the guidelines as general guiding principles of communication and move on."

Harley and Hanaford looked at each other in surprise.

They continued the discussions all day long and by the end of Monday they drafted their statement. The local groups would remain autonomous. For the first five years, the cooperation between groups would be facilitated by a network that included the instantaneous communication system with an annual convention once per year. The EPA's role would be limited to providing a framework for the subsidization of the process with a monitoring role that would determine the amounts of funding that would be allocated as block grants to the local groups.

There was some dissension on this point. Bertrand and Shelski raised the concern that the EPA would impose its own controls of the process through such "strings," which may return the grassroots mitigation to a federally controlled process.

Hanaford and Harley sided with Curtiss. The group turned to Mitch, who could have tied up the point of conflict with a vote that sided with Shelski and Bertrand. He looked at the side of the room at Reverend Tyndall and thought of how he would have resolved this stalemate. He thought that Reverend Tyndall would have stayed out of the medley and directed the panel to take a break. Instead, he asked them all to think outside the box but continue the discussion. Around the table, they discussed and compromised. In the end, they adopted the statement with a provision that would require the EPA to offer funding based on demographics and achievements. The "achievement standards" would not be determined by the EPA. Instead, they would be formulated by the Grassroots Summit Council at the annual meeting.

Mitch drove home with Tyndall and Shelski in the electric vehicle that had been provided by Bertrand. They discussed the odd behavior of Curtiss. Bertrand and Shelski felt that he must have something up his sleeve. Mitch also had a suspicion that he was hiding something.

Reverend Tyndall offered his counsel that they should not focus on Curtiss, but rather on the mutually agreed upon common objectives of the group.

Mitch looked at The Rev and said, "We work toward creating a unified compromise front with at least the majority of the members. If Curtiss does have a hidden agenda, it will come out in the end. If this is the case, you can deal with it through a united front from the remainder of the group. Mitch recalled how effectively that process worked at the geo-engineering science panel deliberations of Door #5.

Then, The Rev turned to Bertrand and Shelski. "I know what you are thinking. Hanaford and Harley have been siding with Curtiss so far. However, if we do uncover a deceitful plot, I think that they both will turn against Curtiss. Hanaford and Harley may be in disagreement with our perspective, but they both have integrity."

Suddenly, the area was rocked with an explosion. The rear window shattered in a glass mosaic spider web and a flash of fire whizzed by Mitch's head. Shelski hit the gas and deftly drove out of the area.

"Goddamn it!" Bertrand barked. "Those vigilantes are at it again." Mitch looked at Tyndall. "Ever since this grassroots process began, there have been reactionaries that feel threatened about their way of life. These are people who cannot adapt to the change of our new reality. Some of them have attacked some of the great projects that the grassroots groups have constructed." Bertrand" sped away out of danger.

The next day started out on a sullen note. They learned that the attack on their vehicle was not isolated. There was a bomb that exploded at a dinner gathering of some of the local representatives. Mitch asked Tyndall if this civil disorder was representative of a society in anarchy. He was relieved to hear that it was not nearly that bad. The law enforcement network of this world was clearly in control. Indeed, security was tightened at Faneuil Hall and each of the participants was provided police escorts to and from the summit that day.

The work of Mitch's Grassroots Planners panel proceeded smoothly throughout the day, as they composed a list of the areas where the local groups could cooperate with each other, including the development of a local technology-sharing process and a network of exchange visits that would encourage local groups to host other community leaders to observe their community's efforts.

At the end of the second day, the entire team met for dinner at the View of the World restaurant in Boston. Curtiss declined the invitation. This was a relief to Mitch (and probably to most of the others).

As it turned out, Curtiss' absence was an opportunity for Shelski, who needed a confidential discussion with the group. "Actually, being here without Curtiss saves us from having a separate after-dinner meeting later. I received an e-mail that was intercepted by my colleague at the EPA, Shelly Rogers, that contained a communication between Curtiss and his own ally within the Agency, Tom Campbell. I have both Shelly's e-mail and the attachment with Curtiss's private e-mail to Campbell. Why don't we read these after our meal so as not to spoil our appetites?"

There was little conversation during their meal. The group was apprehensive about the content of the e-mail. During their after-dinner coffee, Shelski passed around copies.

Dear Dr. Shelski,

I hope that your work at the Grassroots Summit is going well. I need to relay this information to you from my EPA colleague Joanna Simon. No doubt, you have met Curtiss, a reactionary bully who is a known pain in the ass throughout our agency. However, I must warn you that he has a following. In fact, there is a growing schism here between those, like myself, who would like to see the grassroots communities foster their own development with the EPA simply as an "emeritus adviser." You folks have made so much progress without government interference, why would we want to re-establish our efforts with our tired old bureaucratic process? Anyway, on the other side of the fence there are those—led by General Curtiss—who want to rein in your efforts and re-create you as a separate "citizen advocate" division of the EPA.

Read the attached communication between Curtiss and Dr. Tom Campbell, who heads our public outreach program. You may have noticed that Curtiss is masking his difference of perspective with a "gentleman diplomat" facade. Be careful of his tricks. It appears that he is ready to ensnare you with his trap in a way that you will not even notice.
Good luck!
Shelly

The Blue Door (# 7): The Grassroots Solution

PDF Attachment

Hello Tom,

I wanted to brief you on this summit of the whack-a-doodles. First of all, Harley and Hanaford are in our camp. As I expected, Drs. Shelsky and Bertrand are both aligned with the position that the grassroots need to be independent. This Dr. Mitch guy from the unknown time and place worries me. I swear I worked with him on some panel in the past—for God's sake it may even be the future, the way these far-outs are acting—and I vaguely recall being boxed out by him.

So that gives us a 3–3 deadlock when push comes to shove. I have taken your advice and I am being as diplomatic and friendly as I can—I even went along with their votes on the statement of mission and the areas of cooperation. Tomorrow is when I will sneak in our secret provision. They trust me, now. Their guards are down. When I propose the provision that the entire structure be placed under the aegis of EPA's "Concerned Communities Program (CCP)," they will not suspect a thing. None of them know of the fine print that our subsidies to the grassroots communities will be based upon the CCP's requirement that all funding is for a transitional program that transfers to EPA's authority in five years. The action will be viewed as a footnote to our final document and will not be debated. They will all think that it is just throwing a bone to old Curtiss. Then our underground group will work on ways to absorb all of the initiatives under the EPA's structure. If we make the transition gradual enough, the grassroots people will become addicted to their grant funding and will not even know that they have been absorbed under our administrative structure.

So, let me know if you have any suggestions before tomorrow when we will execute our plan.
Curtiss

After everyone had read the e-mail and attachment, there was a period of dead silence at the table until Mitch broke the ice. "Well, that's an eye-opener. You guys were right with your instincts about Curtiss. I guess I was hoping that his accommodating attitude was sincere. Instead, we are dealing with a wolf in sheep's clothing."

Shelski shook his head with dismay. "And I was just starting to trust his diplomatic demeanor. It looks like we will be stalemated tomorrow."

Mitch turned to Reverend Tyndall. "So, Rev, do you have any words of wisdom?"

The Reverend looked at each of them in a calm manner. "You know that I always seek consensus solutions. However, with this dirty pool that Curtiss is playing, total consensus is impossible. There is still time to work an agreement that can be adopted by five out of six of you."

Shelski, Bertrand, and Mitch looked at each other with surprise. Bertrand spoke. "Do you really think that we can get Hanaford and Harley on board?"

Mitch smiled. "Well, if we meet up with them tonight, they may be just as outraged at this subterfuge. Of course, also there is also the risk of one of them alerting Curtiss that we are on to his scheme."

After some discussion, all three agreed that Hanaford and Harley were men of integrity and would judge this scheming to be over the line. Bertrand called them and arranged a meeting at the hotel for later that night. The group started to review how they would trap Curtiss and foil his scheme.

Tyndall noticed that there was someone sneaking up on a man at a nearby table sitting by himself. The stalker reached out toward the man's neck and a flash of blue appeared. A Taser. The unsuspecting man then jerked in a spasm and in a sudden commotion the restaurant was filled with armed police, including a bomb squad, who herded the loner and pulled off his shirt, revealing a bundled vest. They carefully took off the vest while the man was still in a spasm and worked at the vest frantically.

"My God," said Shelski, "what is going on?"

Tyndall, who had seen the episode in its entirety, told them that it appeared that the man sitting at the table was stunned by a Taser and his vest must have been a bomb. "Well, I guess we are out of here, but we might as well wait for the panicking crowd to exit first."

"In the meantime, under the table," Mitch advised. They all kneeled down, anticipating a blast, but it never happened. The police escorted the man through the shrieking crowd.

Mitch jested, "Well, our plan to foil Curtiss almost went up in smoke." They all laughed and quickly exited the cafe.

As they drove to the hotel, they talked about the violent efforts to disrupt their summit. "I'm flabbergasted by the lengths that these reactionaries will go to deter the formation of a union of grassroots organizations," commented Mitch.

Reverend Tyndall responded, "This is a big change to the way our society does business. I believe that this reactionary response is coming from those who would have their vested interests upset. It is normal to expect pushback from those who feel threatened by big changes. The integration of the Down to Earth movement into society's way of doing business should be a goal. The goal should be flexible, but the mission should be immune from compromise. As for the violence from the reactionaries, unfortunately, we have to look at this as collateral damage. I am not sure that I will be around to see the outcome of this summit, but I am hopeful."

Mitch was disturbed by the possibility of Tyndall somehow exiting the scene. He had warned Mitch that he might be passed off to a new guide at some point, but he had become accustomed to the man's calm presence.

When they returned to the hotel, Harley and Hanaford were waiting in the lobby. Hanaford asked, "What's up? You guys sounded serious. Did you hear about the bombing attempt at the cafe?"

Mitch responded gloomily, "We were there. Pretty scary."

Harley joined the group. "Are we waiting for Curtiss?"

Mitch shook his head. "No. This is just us; we'll explain. Let's find a room."

The concierge took them to a small conference room. Mitch, Bertrand, and Shelski shared their findings about the plot schemed by Curtiss. Mitch was apprehensive about how they would react. It was certainly possible that one of them would have some loyalty to Curtiss and expose their awareness of his plot to him. The room was silent for a long pause after they finished their briefing.

Hanaford reacted first. "That son of a bitch! It is one thing to challenge the wisdom of letting the grassroots groups go independent of the EPA. It is another thing to hatch a clandestine scheme. I feel that we are being duped by him and his henchman at the EPA."

Harley added to the fire. "I feel like opening the meeting tomorrow with a motion to hang the bastard." There were some chuckles at the table.

Mitch looked at Reverend Tyndall, wondering how he would handle this situation. Then Mitch addressed the group. "Okay, guys, back to reality. It is actually very simple what we need to do, assuming that you are on board. Our main task tomorrow will be to hammer out an agreement for ongoing communications among the local grassroots groups to coordinate mutual aid. I believe that the four of us can come to a compromise proposal. Then—probably in the late afternoon—we will review the digital record of all of our meetings and compose our final agreement document. I suspect that Curtiss will lay low until the very end. Then he will take advantage of our weariness over the last all-day marathon deliberations and will add on his rider for the entire program to fall under EPA's CCP program. If you are all on board, we will simply pass a motion to reject this amendment and follow with a motion to approve the entire package and adjourn. What do you all think?"

"You forgot something," said Harley. "The motion to hang Curtiss." They all laughed. Then they agreed to Mitch's strategy.

Mitch turned to Reverend Tyndall. "I know that you are always a full-consensus guy. In this case, Rev, I don't think that we have the time to get this package wrapped up for the full summit with a lying scoundrel undermining our process. We have to box him out."

Tyndall chimed in. "There is a time for accommodation, and a time for confrontation. Unless that e-mail was a fraud, you need to move past this deceitful intransigent man and move on. You will know if the scheme is true if Curtiss makes the motion to put the grassroots program under the aegis of the EPA's CCP program. While you were discussing this matter, I did some research on my brain pad, with a search filter for verification by corroborating documents. It is very clear that the EPA's Concerned Communities Program was intended as a start-up for local initiatives that would be folded fully into the Agency's administrative structure within five years. Very few know about this obscure program. In fact, it has worked very well in the manner that it was intended—to cultivate ideas from the community groups who started the work and fold their ideas into the EPA's control. The intent, however, is for the community groups to know up front what is happening."

Reverend Tyndall looked at Mitch and patted him on the shoulder. "Mitch's approach is the only way for you to deal with this scheme. I have seen him in action before and he'll know how to box out Curtiss. I also admire both of you." Reverend Tyndall looked at Harley and Hanaford. "You have had your reservations all along, yet you were open minded enough to discuss and compromise your perspectives. I believe that the discussions, your disagreements, and the give and take, added strength to your findings. You should all be proud for standing up to this subterfuge with strength and integrity."

Bertrand seemed entranced by Reverend Tyndall. "Who are you, anyway?" he said. "All along you have been a silent observer, but when you speak, you are so inspirational."

Mitch followed up. "You see his white clergy collar. That symbolizes wisdom, in my eyes." Tyndall blushed but bowed to him for his compliment.

Tyndall went on. "I have been following this Down to Earth movement and it is a breakthrough in the gridlock of other approaches to resolve this climate crisis. I firmly believe that you are on the right track. The people take the lead at the low rung of the ladder and feed their solutions from the bottom up. That is the best hope that we have. You gentlemen have made a profound contribution—even if history never recognizes your achievement, you have set the stage for this wonderful process to spread throughout the nation and the world. I bow to you all." And he did.

With that, they adjourned for the evening. All of them seemed renewed in their harmony. As they exited, Bertrand whispered to Mitch, "Who is that guy? He seems almost mystical."

Mitch responded, "I have known The Rev for a bit longer than you. He does have a charismatic glow and it is genuine. His 'wisdom' arises from his Ph.D. in environmental ethics and years of experience in mediation with groups like ours. If he seems otherworldly to us it is simply that he is a man with an advanced level of wisdom. He is very human and quite an asset."

The Blue Door (# 7): The Grassroots Solution

Tyndall, who had caught up to them, said from behind, "So who are you calling an ass?" The three of them laughed.

Before turning in for the night, Mitch invited Reverend Tyndall over for an elixir. He brought out two orange drinks from the refrigerator. They watched a bit of the World Series where the Chicago Cubs were playing the New York Yankees. The score was 3–3 in the seventh inning. After a while, Tyndall turned off the set and groaned. "I can't take it. I've been a Cubs fan for so long, I'd just as soon read about it tomorrow."

Mitch laughed. "So, you're a Cubs fanatic. I had a friend in another dimension that was just as crazy about the Cubbies as you are. Wait a minute! I thought that the Cubs won the World Series somewhere around 2017."

The Rev responded. "Remember, you are in a different dimension."

Mitch took a sip from his orange elixir. "I remember a time when we had nightcaps with beer or wine. Now it's an elixir. This drink has grown on me. It's more calming than tea." He then turned serious. "Rev, I am sensing that we are approaching the time when we will move on to another era. I'm hoping that it won't be another committee. I am a man of action; I'm just not cut out for running committees."

Reverend Tyndall loosened his collar. "Very soon you do move on to a different era of Door # 7. You may be picking up from your intuition that things will be different on our next stop. Perhaps you will not be disappointed with it. In the meantime, feel good about what you have achieved. Just as with the great volcano panel, you orchestrated this group like a maestro. I am sure that you will wrap things up as brilliantly as you have done in the last two days. I am proud of how you have grown." The Rev let out a loud belch. "Well, excuse *me*."

Mitch laughed. "I wish the others could have witnessed that to see that you are all too human. Some of them look at you as a god."

"Mitch, I do have my flaws. You can probably see that I am a loner. The few times I've had the opportunity to participate in these journeys, I do my best to tap into my skills and experience to mediate groups like this. It does seem very bureaucratic and tedious—especially to folks like you who would rather be out in the field. But I think that you will see later in this evening's journey how important it is to set a solid foundation for a vibrant movement that is coming together before your eyes." Reverend Tyndall drained his cup of the elixir. "And now it is time for me to turn in for the night. Thanks for the nightcap, Mitch. Get a good rest for tomorrow." Tyndall left the room.

The next day, the group knocked out a recommendation for a nuts-and-bolts proposal to maintain communications among grassroots groups. This included a website, a chat room (which would schedule dates for events), an annual meeting among all groups, timeframes for regional meetings in-between the annual national meeting, and an outreach to other grassroots organization worldwide. Shelski proposed a creative activity that would have each group conduct its own inventory of carbon emissions in their communities and to track the reductions on an annual basis. All of these ideas were unanimously accepted by the group. Before dismissing for lunch, Mitch reminded the group that they needed to spend some time analyzing the audio cloud recordings of their three days of deliberations. The transcription program would then take the digital file and transpose it into a hard copy document that would become their report. He thought this was a very cool technology—all of their discussions were captured on a digital cloud and with their cloud-access readers, they were able to dissect and highlight sections of their proceedings, flag the items that led to their recommendations, and send them off for automated compilation to produce a report.

Mitch was ready to break for lunch when Curtiss threw his expected bombshell on the table. "I have one last item that we should consider as a rider to all of our proceedings. It is really very minor, but I would like to see our group accept this resource as part of our recommendations. The EPA has a program that perfectly fits our function. It is called the Concerned Communities Program (CCP). It provides the Agency's vast resources to assist community initiatives and foster their growth. It can actually be very helpful in providing very specific milestones and the funding and technical support needed to achieve them. It can be used as a basis for applying for the annual block grants that will be distributed to the local organizations. I know that this is a no-brainer for the group, so I will make a motion that our report stipulates an affiliation of the national grassroots network with CCP."

This motion was followed by stony silence among the other members.

Mitch could sense the confusion in Curtiss at this dead-quiet and the temperature rising from the others. He spoke before someone lit the match of acrimony. "I congratulate you on your initiative to secure resources and technical support from the EPA, General Curtiss, but I frankly think that we are all wiped out from the long morning. Why don't we all think about this idea over lunch? We can take up your motion first thing this afternoon." All agreed and they broke for lunch. On their way out of the room, Mitch whispered to the others, after Curtiss had left, "That was a very crafty move by Curtiss. Catch us off guard when we were most vulnerable due to our collective mental exhaustion. Fortunately, we know better. We shouldn't dine together, as he may see us together and suspect that we know of his scheme. Let's, instead, communicate on our Xeri transmitters on the confidential channel."

During their lunch break, Mitch suggested a strategy for a very simple resolution to Curtiss's motion. The entire group concurred that his idea was a very clever approach.

The afternoon session started with the six panelists in a more refreshed disposition. Curtiss looked confident. Mitch and the others were relaxed. Mitch opened the afternoon session. "Okay, folks, there is a motion on the floor. Do I hear a second?" Shelski promptly seconded the motion. When Mitch called for a discussion, everyone was quiet for a moment.

Then Curtiss spoke. "Well, I think that this is a minor matter that will be a benefit for the grassroots network across the nation. The EPA's CCP program has been running for ten years now and has supported very productive local initiatives. There are numerous benefits. Take a moment to review some of the achievements of local communities who enrolled in CCP in areas ranging from local water pollution problems to protection of habitats, among countless others." Curtiss handed around a one-page glossy program summary that was printed by the EPA. What should have taken a few moments stretched out to an uncomfortable five-minute pause in the conversation.

Finally, he looked around the table with a confused demeanor. "So, what do you all think?"

Bertrand responded, "I move the question."

Curtiss was pleased that they had dismissed any discussion and were moving toward a unanimous approval of his rider. Mitch called for a vote. The motion failed 5–1. Curtiss looked shocked. "What the hell is going on here? You all decided your vote without the courtesy of your thoughts on why you are opposing it. I would like to know your reasons!"

Harley responded, "I think that we can answer your question with a motion of my own. I move that we request the EPA's assistance from the Concerned Communities Program (CCP) but without the mandatory affiliation and five-year fold-in to EPA's administrative structure." Bertrand seconded the motion. Mitch called for discussion.

Curtiss was visibly agitated. "You can't do that! The program requires guidance by the EPA and you should not be threatened by it. We are only trying to help you folks. There is nothing in our program that requires dissolution of your group."

Bertrand responded, "Actually, I have the regulation that created the CCP. I will quote you the salient language. 'Participants that have been accepted into the Concerned Communities Program will have a six-month period to consider the ongoing financial and technical services benefits of the program. At the end of the trial period, the EPA will gradually assume authority over the program of the community group. The EPA will then have full authority to administer the program and consider the input of the recipient community under the EPA's direction for funding and services.' "

Curtiss responded in a thunder, "You have this all wrong! The formalities of the language are only that—formalities to satisfy the EPA's sanctioning of funds and services. With your motion, Harley, you are looking a gift horse in the mouth and then kicking it in the teeth!"

The general's attitude was expected. This all helped to cement their decision even more. Each one of them also knew this opened up a whole new can of worms.

Harley chimed in with a calm voice. "Actually, Curtiss, you are incorrect. Section A (2) (i) of the regulation states that the community organization applying for CCP assistance may request that the EPA provide said funding and services with a waiver of their mandatory affiliation, which will allow the recipient community to receive this aid without formal assumption of the program. That is the essence of my motion."

"But such a minor detail! You have six months to decide on that option, why institute it from the start? Don't you trust the EPA?"

Shelsky replied, "Curtiss, we don't trust *you*. We know that the EPA is split on the whole idea of grassroots programs and that you are firmly in the camp of those advocating that the EPA take control of the entire process."

The table was quiet for a period of time. Curtiss appeared to be composing himself. "It seems that I am the odd man out in this discussion. I don't know how you contrived to obtain this information but I am with an agency that has provided innumerable services, which I believe could have helped your program. Since I am clearly in the minority, I voice my opposition to your applying for the exemption of the mandatory affiliation. This is insulting. You are acting as if I have concocted some conspiracy. Hanaford and Harley, you have been with me all along on the issues of the last three days. I do not know what I did to lose your trust, Harley. In any event, I think that we should take a break, cool down, and return to the table to reconsider my original motion and rescind the approval and replace it with one that makes more sense. Let's reason with each other instead of antagonizing Hanaford and myself into a forced minority."

Hanaford responded, "Well, actually, you have not been up front with us, Curtiss. We have received a copy of an e-mail between you and Dr. Tom Campbell, which you sent two days ago. You revealed your subterfuge to co-opt us into EPA and wrongly assumed that Harley and I were in your camp. I stand by the approved stipulation."

Curtiss stood up and yelled out once again. "This is an outrage!" His face was flushed red. "You can't take a vote without me. Our guidelines required a unanimous consensus on the full package. If you go forward with this vote, I will walk out of here."

"Well, Curtiss," Mitch responded, "first of all, the guidelines of our conference list you as an ex-officio member. Up until now, we have certainly given you the courtesy of your vote and continue to do so with this motion. But the fact is, if you walk out on us, it really doesn't matter, because your vote doesn't count, as an ex-officio member of the panel. So I roll with the vote. All in favor—"

"You can all go to hell!" Curtiss kicked his chair and walked out of the room.

Bertand looked at Mitch, "What now?"

"Now that the vote has passed . . ." Mitch said, "We move on. Curtiss recorded his opposition to the motion after it was seconded. So we will record his negative vote and his actions."

The remaining panel put together their report by means of the instantaneous auto-cloud transcriber to compose their recommendation package. Mitch orchestrated a unanimous acceptance of the report and transmitted their findings to the full summit council. He then suggested that they all walk over to the Red Sox pub across the street for elixirs and burgers.

"Aye-aye, Captain!" they all boomed.

Harley summed up the group's sentiment. "Hail to our chief."

Mitch prepared to conclude the meeting. "So we will adj—"

"Before we close your proceedings," Reverend Tyndall interjected, "I would like to take a moment to congratulate you on your efficiency, your diplomacy, and your professionalism. I have been a process observer for the last three days, and I must say that I have been very impressed. I predict that the good work of your Grassroots Planners team will be setting the foundation for the most significant social change in history. I tip my hat to you all!" Then Tyndall did just that. He took off his cowboy hat, folded his hands, and bowed in a prayerful acknowledgment of their work.

At the pub, they enjoyed light hearted conversation. Tyndall actually opened up, talking about the Cubs taking the World Series the previous night. He offered a toast. "To the Cubbies. The jinx of the goat is finally over."[59]

They all raised their lemon elixirs and in unison said, "Here, here! Hail to the Cubs!" Then they discussed their maneuvering around Curtiss.

Mitch surprised the others with his reflection. "I actually feel bad for him. He must be humiliated. I know that he planned subterfuge, but I for one can forgive him. He was just using his 'alpha male' aggressive strategy to get his way. But at his core, he wanted these local activist groups to feed good ideas to the EPA—just on his own terms."

Harley added, "Too bad that we couldn't have him at our table to end on a positive note."

Reverend Tyndall responded, "Well, if you would like, he is waiting in the next room sitting alone with a beer."

They all decided to join Curtiss. Mitch led the way and surprised the man with an outstretched hand. "No hard feelings, Curtiss. You were a valuable member of the panel. During the sticking point, you proved to be a worthy opponent."

Curtiss blushed and shook his head. "I guess I was a bit underhanded. When Reverend Tyndall asked me to be here, I actually hoped that you would join me so that we could all make amends."

For a moment the four of them looked at Tyndall.

Mitch smiled with an uplifting thought. *What a classy guy. I am certain that the others are thinking the same at this moment. The Rev—always the super mediator.*

---

59 In 1908, the last time the Chicago Cubs won the World Series, a disgruntled fan who was denied entry to the game because he tried to bring his goat into the stadium, He put a curse on the Cubs that they would never win the World Series again

Reverend Tyndall then offered another toast. "To the Cubbies! May they live long and prosper." All laughed and they enjoyed an evening of levity.

On their way out of the pub, Shelsky patted Tyndall on the back. Bertrand piped in, "You are amazing. Your mere presence at the meeting seemed to keep us on an even keel."

Harley quickly chimed in, "You are quite a mediator and a genuinely good guy."

Tyndall responded in a Texas drawl that were to be his last words. "Aw, shucks, guys…"

Fifteen minutes before they were to leave the pub, in the dark recesses of the alley they would be passing, a hooded figure took refuge in the darkness. He took his 9mm Glock out of his pocket and slowly put the magazine in place. He looked left, then right, and silently retreated behind a dark blue trash bin as the lights from a passing car exposed him in a flash of light. He waited a moment and then pulled the slide back, revealing a shiny hollow point bullet. The hooded person let the slide snap shut, loading the weapon, checked the laser sight, turning it on and off and waited in the shadows breathing heavily as his target—the man in the clergy collar—came into range. His target, Reverend Tyndall, was now in the telescope sight of his Glock. As Tyndall was giving his "aw shucks, guys" comic delivery, the hooded figure pulled the trigger. *BLAM!* Suddenly the silence of the street was broken by a single gunshot. Tyndall went down, with a bloodstain quickly spreading across his shirt.

Mitch gasped. The cacophony around him was drowned out as he receded into his own mind. *Oh my God, no.* He knelt down to where Tyndall was lying. His priestly white collar and his bolo tie were covered with red.

Mitch felt a powerful pang of sorrow knowing that his colorful guide was fading away. "Hang on man, please. You can't leave me now." He wondered why Tyndall was a target. Was he singled out or was it a random shooting?

He looked up to Mitch with dying eyes. "No matter, my friend. There are other dimensions than this. Let it Be." Then Reverend Tyndall slipped into the long dark beyond.

Mitch felt an array of emotions and despair on losing his friend. The practical loss was the worry that he no longer had a guide. Was he stuck in this dimension for the rest of his days? Would he lose Leah and Susannah forever, too? But the profound loss was that he would never see his mentor—and friend—again.

As if the ethereal world beyond his grasp had read his thoughts, from the sky there came the spectacle of a large bubble appearing to the side of the chaotic scene of the shooting. The bubble was the color of a rainbow, but dominated by blue. No one else seemed to notice its appearance. Inside the bubble was a stunning lady, dressed in a flowing white gown, with long red hair and the face of an angel. Her eyes had gold irises. She beckoned Mitch inside and he was suddenly enshrouded in the indigo cloud. He intuitively knew that this was the new guide. As they floated away, he saw Bertrand

and Shelski looking up with awe at the bubble floating away from them. Mitch looked up at her and asked if she was her new guide. She smiled at him and started signing her words.

Oh, dear, she is a deaf-mute. Her first words were signed with individual letters.

"I am Gabriella. I am your guide for the last leg of your journey."

Mitch signed back, "I guess you must know I can sign and respond to your words. My younger sister was a deaf-mute. I am pleased to meet you, but I am very shaken and sad."

Gabriella laid her hand on Mitch's face. "Let it Be," she signed.

Mitch took a deep breath. "So, here we are," he signed back. "Now where are we going?"

April 22, 2152

As if answering his question, the bubble floated on to a complex of futuristic buildings. The indigo blue bubble dissolved as they landed near a strikingly blue building All of the windows had the green tint of the super-efficient energy-trapping windows. All of the buildings of this urban hub were shaded with a glowing blue that contrasted with the green insulated windows. Although utilitarian in design, the buildings were surreal. A road sign said that they were in the City of Technology, a makeover of Albany, NY, which Mitch had visited years ago.

He signed to Gabriella, "I can't believe that this is Albany. It all looks so strikingly clean and high-tech."

She signed back, "Most of the cities of this grassroots era were either vastly retrofitted or rebuilt altogether. Albany is one of the former."

She guided Mitch up the stairs to a building that was identified by a glowing, light blue sign that appeared in the thin air at the top of the steps: The Museum of the Grassroots Movement: Floors 1–3. The words dissolved and were replaced by another set of words: The Climate Monitoring Center: Floors 4–6.

That is very cool—like those clocks that displayed time and date in the air above the timepiece, back in my dimension.

He turned to Gabriella and signed, "What year are we in?"

Gabriella pointed to the same air space that had identified the building. The ethereal now read April 22, 2152. 10:00 a.m. She then guided him to the museum on the first floor. Mitch was awestruck by sixteen different displays that were stunning in their depiction of what had transpired over the past 130 years. The visuals were instructive, from transportation models to displays of the new agricultural centers. Equally inspiring were the ethereal words that appeared in front of each display. He took notice of the display at station 8, which was a marble statue of his guide and friend Reverend Tyndall laying in the road where he'd left him. After he read the sign at station 8, he realized that Tyndall became a symbolic martyr for this new age.

Gabriella signed, "Let's take some time to orient you to our progress. We'll spend a couple of hours here in the museum. Then we will go up to the Climate Monitoring Center." Mitch stopped at each station. Although the graphic presentations of the world's story were almost magical, it was the information in the explanations of the grassroots history that had him entranced. When he finally finished viewing the displays, he needed to sit down as his mind was swirling with the information about the displays and what had transpired since 2034. This was very similar to the exhibits on geo-engineering through Door #5. All of the holographic explanations were churning through his head. Mitch sat on a couch, exhausted. Gabriella gracefully sat on the chair next to him. He mentally re-read what he had just learned:

- *The Grassroots Summit of 2034 was a catalyst. The movement adopted the name 'Down to Earth.' The summit integrated a combination of local ideas and no-strings-attached funding and technical support by the federal government, innumerable green projects blossomed and were shared among the community groups.*
- *By 2040, forty-eight out of fifty states had provided funding and support for local Down to Earth projects.*
- *New local green utilities started by citizen groups increased the market incentive for renewable energy. Nationwide, the amount of "green energy" in the national grid quadrupled the amount of non-fossil fuel energy.*
- *A coalition of Down to Earth groups proposed a "smart grid[60]" coalition with utility companies that drew seed funding from the national "down to earth clearinghouse" and a national subsidy from the federal government for implementation. The national smart grid was constructed in 2080 and decreased emissions from home electricity by 17 percent.*
- *The momentum of the Down to Earth movement sparked creative alternatives for doing business at all levels.*
- *A council of local, regional, and state governments followed the lead of Down to Earth and in forty states they have implemented pilot projects of alternative transportation.*
- *The automobile industry jumped on the bandwagon and all of the major car companies created a plan to convert to electric, hydrogen cell, and other alternatives by 2100. All of these companies signed a pledge to provide 90–100 percent of these vehicles to the public by that year at a reasonable price. It was anticipated that market pressures would phase out the gasoline automobile by the end of the*

---

60 See http://energy.gov/oe/services/technology-development/smart-grid

- twenty-second century. As it turned out, gasoline-powered engines were out of the picture be 2140.
- In the "old cities" of the U.S., the resistance from the stubborn reactionary groups became more and more dangerous. Under a marble statue of Tyndall, was an inscription that told the story of his martyrdom during the first Down to Earth summit of 2034.
- The grassroots centers started to build out into communities of their own, such as in the Town of Silver Creek, south of Buffalo, which became the "Technology and Monitoring Community." The nation's economy—business and industry of the new era—also found it infinitely more profitable to abandon the old ways and integrate themselves with the Grassroots Cities.
- By 2040, the resistance to the Down to Earth movement had faded. In that year, a coalition of philanthropists provided a $100 trillion pledge to the Down to Earth clearinghouse to distribute to local projects at their discretion. The coalition also pledged to continue their donations on an annual basis. This funding seemed to send a signal to those resistant to the societal changes that "If you can't beat them, join them."
- Local farming and organic "small carbon footprint" agriculture provided significant competition to Big Ag, inducing them to adopt green farming practices.
- In 2048, Canada and Mexico joined the U.S. national clearinghouse, based on the success in spurring innovative technological solutions to climate change that had spawned a new green industry. This was another economic catalyst that spurred national economic growth.
- A key element of the Down to Earth coalition is climate-emissions monitoring. Each community in the network takes an annual inventory of its emissions from vehicles, buildings, and industry and tracks emission reductions over time, factoring in the improvements and upgrades.
- A study was conducted in 2090 to determine the effects of the Down to Earth movement from 2034–2080. Amazingly, the U.S. had reduced its total carbon emissions in the fifty-six-year period by 70 percent. By 2087, the reduction, relative to 1990 levels, was measured at 85 percent with the further maturing of these efforts (such as the completion and implementation of the national smart-energy grid and the phasing out of fossil fuel transportation). Additionally, an innovative technology—the Ambient Air Removal Technology—was deployed in 2045. This supplemented the carbon emission reductions from single-point industrial discharges and sucked the CO_2 out of the atmosphere. Many looked at this as the cleansing of the sins of the Industrial Revolution.

- *The astounding success of the U.S. Down to Earth movement sparked a European initiative in 2048. They convened a series of grassroots summits and created a European Green Clearinghouse based on the U.S. model.*
- *In 2050, China, Vietnam, Australia, and Japan created an Asian grassroots network, with a strong local and regional component of alternative green technologies. Incredibly, the Chinese government encouraged these decentralized efforts to reduce the air pollution that they had not been able to achieve through top-down efforts. In 2051, India, Malaysia, and the Philippines joined the coalition.*
- *In 2081, a coalition that included the U.S., European, and Asian networks sponsored a series of "Green Corps" (funded by a coalition of philanthropists) that created their own sustainability priorities. These teams were then sent to third-world countries to help foster development with low carbon technologies. The first dispatch—aimed at reducing black carbon emissions[61] by 75 percent by the year 2100—sent teams to villages to convert their traditional indoor fire heating to clean cook stoves that vastly reduced this source of CO_2 (and made their homes safe from carbon monoxide poisoning). Under discussion is the encouragement of national smart grids to power local villages in the undeveloped world with low carbon renewable fuel sources, solar power and wind power. However, this international coalition decided to promote this as a series of subsidies and Green Corps support that would be available to them only if the local governments requested their help.*
- *Worldwide, the reductions of emissions lagged behind that of the U.S., but the fact that there was a 55 percent global decrease by 2075 showed progress. But, violent reactionary militias in the undeveloped world and the developing nations exhibited the stiff resistance to the Down to Earth efforts that continues right to the present.*
- *In many nations, the federal governments offered to take over the efforts to allow uniformity and oversight over these very successful efforts. In all cases, the individual Down to Earth organizations declined, but offered the suggestion that the governments follow the U.S. model to provide subsidies to the networks, with no strings attached. Such subsidies would be revenue neutral, based on the savings and additional job growth that has been realized from the new grassroots technologies.*

Mitch turned to Gabriella and signed, "I am hungry and exhausted. Can we take a break before going up to the Climate Monitoring Center?"

---

61 See https://www3.epa.gov/blackcarbon/basic.html

She smiled at him and opened her hands. An indigo bubble appeared in front of them. She waved her hand in front of the bubble and a menu appeared. Mitch ordered his lunch, which he washed down with a grape elixir.

Gabriella then pointed to the bubble and signed, "Sleep. No one will see the bubble so you will not be conspicuous."

Mitch signed back, "Thank you" and entered. He sat down inside a cloud in the bubble and immediately fell into a one-hour nap.

Mitch awoke with a start and wondered where he was. He was bathed with the glow of blue around him. Then he remembered. He and Gabriella exited the bubble and walked up to the fourth floor to the Climate Monitoring Center. The open room was expansive with digital image projections of the earth's surface that covered the walls. The workers in the room were intensely busy—it appeared that they were in a world of their own. The energy of the room was electric. They were approached by a woman who was clearly in charge of the operation. On her way to greet them, she periodically stopped, checked the work at individual tables, and gave directions. As she approached, Mitch was taken off guard. It was Leah-Ann—his descendant.

I should have known this was coming. He turned to Gabriella and signed, "How do I handle this? This is the first time that I will interact with my descendant."

She signed back, "Let it Be. She knows nothing about your appearance from a different time and a different dimension. I will be in the next room to allow you to speak to Leah-Ann one-on-one."

Leah-Ann came over and introduced herself. She looked a long time at Mitch. "This is very odd," she said. "You look awfully familiar. I can almost picture you with my family at Thanksgiving last year."

Mitch recalled what Jake and Janey had told him. Leah-Ann cannot see between dimensions, although she might have some intuition. He also recalled Jake and Janey's warning not to cross the lines between different dimensions.

So he responded with a casual smile. "Well, it must be some unexplained memory—what we call 'déjà vu' in the place where I call home."

Leah-Ann still seemed disturbed. "But your name. Mitch Warner. That was the name of my ancestor from the early 2000s."

Mitch was quick on his feet. "Coincidences can be quite strange."

Leah-Ann shook her head and accepted that.

Mitch, on the other hand, felt a range of emotions, proud of the leading position of his descendant, sad that he could not reveal to her who he was, and a profound love for the distant bud of his family tree.

"Okay," said Leah-Ann, "I am told that you come from a different place that I am not supposed to ask about. I am to put you to work as my assistant. I direct the world's

Climate Watch Program. We have a dynamic team here at the center. Most of the detailed work is done at numerous stations on the floors above us. This large room is the central command, where we monitor the overall state of the climate worldwide. Come. Let me take you to the interactive map projection."

The world map was enormous, covering the entire wall of the room. It was peppered with different-colored pinpoint lights over the entire world map.

"I know that you are going to explain the meaning of the light points, but I assume that the red lights are problem areas."

Gradually, Mitch fell into his active participant role, and the workings of the display became familiar to him.

He spoke, again, to Leah-Ann. "I want to see the big picture, but I also want to take a look at the specific locations you are taking me to."

Leah-Ann looked at Mitch with a bit of surprise. "So, you know of your assignment?"

"Well, I know in a general sense what we are doing, but you will have to fill me in on the specifics."

"Hold on, Mitch, I will get to the locations where we will journey, but bear with me as I explain the big picture."

She pointed above the map to a display that gave the current-moment data of the world's greenhouse gas situation. It showed that the emissions rate worldwide was 7 gigatonnes (GTs) per year, compared to 35 GT per year in 1990. The total amount of CO_2 in the atmosphere was 1500 GT compared to 1650 GT[62] in 2020. The fact that there was a decrease in the CO_2 in the atmosphere was significant. It would take a few centuries for the atmosphere to come to equilibrium with pre-industrial revolution levels. The average temperature was -0.2 degrees Celsius lower than 1990.

Mitch asked. "I am surprised that the atmospheric pool of CO_2 gone down. I would think that the atmospheric concentration would take centuries to decrease, given the persistence of the gas in the atmosphere."

Leah smiled. "Good question, Mitch. The concentration in the atmosphere is now 390 ppm.[63] We have had success, worldwide, with the Ambient Air Removal Technology (AART) that sucks CO_2 out of the atmosphere. So, by reducing our emissions by 80 percent, the 1,000,000 AART stations worldwide are now slowly reducing the 300 year post Industrial Revolution pool that remains in the atmosphere. Mitch added, "I understand that—the bathtub analogy. It isn't until we pull our emissions rate to a 100 percent cut

---

62 See http://energy.gov/oe/services/technology-development/smart-grid.
63 The concentration in the atmosphere is the unit expressing the amount of CO_2 in the atmosphere that is available to cause warming. In contrast, the pool of CO_2, mentioned earlier as 1500 GT, refers to the mass of CO_2 that is stored in the atmosphere and how long it resides there. See http://globecarboncycle.unh.edu/CarbonPoolsFluxes.shtml

relative to 1990 levels that the total gigatonnes will start falling. But your AART stations have accelerated that removal."

Leah-Ann laughed. "We all grew up with the bathtub analogy. The faucet is still running, but at a much slower flow. When it is nearly stopped flowing with greenhouses gases, the tub level may start to decrease by evaporation and slow leaks from the drain. But you know all of this, Grandpa."

Mitch became alarmed. *She knows!*

Then Leah-Ann laughed. "Sorry about that. We have a custom here at the Center of referring to wise elders as 'Grandpa.'"

Mitch flushed with relief, "No offense taken."

Leah-Ann continued, "So we have reduced our total emissions by 75 percent relative to 1990 levels. It will be hard to get much lower than this—the big emitters in the developed world are reined in, though we certainly can do more. What we have left is the harder to control: emissions from the undeveloped and developing world. However, the question remains: do we even <u>need</u> to do much more? Is the lower emissions rate in the U.S. sufficient in the turning off of the faucet? Will the total CO_2 in the atmosphere come down to safe levels with the AART stations countering our current emissions? That is why the work at the Climate Center is so important. Yes, the onslaught of climate change impacts started decreasing in the 2070s and they are almost gone. The loss of Bangladesh, the Oceanic nations and many of our coastal cities are a sad memorial to the climate emergency of the past century. Yet, we are ultra-cautious to make sure that our battle to reduce carbon emissions continues—particularly in third-world nations that could develop with the same uncontrolled greenhouse gas emissions. As you will see, we have a robust climate-monitoring network with over one million emissions-tracking stations worldwide, and it is so sensitive that a small 'burp' will be picked up. Come. Let me show you one of the places that we will visit."

Leah-Ann zoomed into an area of West Africa; the region now filled the wall.

"A few days ago, our sensors picked up a new emission of CO_2 in this region that was clearly abnormal. We tracked it down to a large fire and suspect there is now a massive burn-down of a rain forest. Nigeria had previously agreed to eliminate this practice and plant new trees. We contacted their government, who verified that there was a burn-down by the militias who are fighting our control efforts. The Nigerian Army has now put out the fires."

After a few keystrokes, the burn area was highlighted in yellow. "In only two days, the emissions have started to subside since the fires went out. Now our job is to visit the local conservation corps to advise them on reforestation and to discuss increases in security against the militias. We will be visiting the Nigeria region last."

Before Mitch could edge in a question, Leah-Ann zoomed in to another region, this time in the U.S. "This is an area in western New York near Buffalo, where there is a multitude of wind farms. The winds off of Lake Erie make this an ideal location for maximizing this green energy source. Let me show you the CO_2 emissions rate today compared to 2034."

With a few entries on her pad, the area lit up orange.

"Our emissions rate of this region in 2034 was 17 kilotons per year from all sources of CO_2 emissions. We saw western New York as a prime area for doing more and our grassroots groups in this area formed utilities of their own that increased wind power and decreased fossil fuel power use. In 130 years, the regional greenhouse gas emissions have been reduced to 3 kilotons per year. Down by 80 percent! We think that we can do better in that area by adding more efficient turbines and tweaking the Robert Moses power-generating facility by tapping more of the river flow from Niagara Falls for increased hydropower. We will be meeting the Down to Earthers to see what can be done to decrease western New York's emissions down to 1 kilotons per year."

She then zoomed in to another region of the globe, bringing a large section of the Midwest into focus. "In 2034, we gave up on our breadbasket. The region virtually dried up, due to the warming climate. Then, new drought-resistant crops and other technologies started a rebirth of agriculture in the Midwest. As farming rose from the ashes of droughts, we aimed to reduce carbon emissions rates from agriculture by 50 percent (relative to 1990 levels) by 2100 using local organic farming methods that would employ unconventional techniques to reignite agriculture, using locally scaled production methods. In other words, 'smart dirt farming,' as we call it. Unfortunately, this is one area where we did not meet expectations. We will be going down there to meet with a consortium of local farmers to see what we can do to get the farming projects on a better track. It has been hard for them to adjust to using organic fertilizers, but the reduction of nitrogen additives did decrease the greenhouse emissions of nitrogen oxides from this region, but only by 20 percent. Although our main goal has been to boost productivity, the reduction of nitrogen oxides was the side benefit. It was amazing that we were able to bring any level of agriculture back to this farm-dead region."

Mitch asked, "Earlier you told us that the major global challenge is in the developing world. So why are these U.S. projects to decrease emissions toward 100 percent worthwhile? It seems to me that we should be focusing all our efforts in the third world."

"Good point, Mitch. First off, the bulk of the efforts of the worldwide grassroots network *are* in the developing world. However, it is a tough nut to crack, as it involves cultural behaviors. So, the U.S. and other developed nations agreed to keep chipping away toward 100 percent of our own emissions to keep the worldwide reduction process going, while we work with the developing world on a long-term change in their industrialization behavior. Also, the fact that we keep moving forward sets a good example."

Leah-Ann then took Mitch to lunch in the Climate Center's cafeteria. He was chomping at the bit for information. "This is all pretty damn good. However, with all of your success, is that really enough? The climate change that is coming—even if you exceed your greatest expectations—is already built into the climate system. I am aware of some of the climate impacts, such as the loss of Oceania, Bangladesh, and parts of Florida to the rising seas, the heat waves, and of course the challenge of farming in a now-arid Midwest. Do you feel that we are turning the corner or just holding our fingers in the dike?"

Leah-Ann smiled at Mitch. "That is the reason why we have such an extensive and sophisticated climate-monitoring system here at the Center. We have our finger on the pulse of nearly all major emissions sources. But we also have modeling of how our climate system is reacting to the lowered emissions. We are able to forecast the effects on all impact parameters. For example, the modeling of sea level rise in 2050 predicted the loss of Bangladesh, Oceania, and many coastlines due to inundation of the sea by 2100. All of that happened by 2070. In 2120, our modeling predicted that the loss of further land areas to sea-level rise that was forecasted in 2050 would be avoided. That has been the case. This same trend has held for other parameters, such as the gradual decline in the number of heat waves, lessening of droughts, and the cessation of the melting of glaciers and permafrost. All of the parameters are on a downward trend from the peak of 2050, and all of this has been predicted by computer modeling from the past hundred years. Also, bear in mind the effects of the AART stations in lowering the CO_2 loading in the atmosphere—we monitor this as well. So, yes, I believe that we are finally getting a handle on restoring climate toward natural conditions, but I would like to see the world reduce the greenhouse gas emissions another 10 to 20 percent or so. I am constantly thinking about my great-times-six grandfather and the long-term consequence of the efforts of his Grassroots Planners panel back in 2034."

She then came over to Mitch and touched him on the arm. "Even if you are not my ancestor, you still remind me of what he would have been like."

He bit back his tears and quickly launched another question at Leah-Ann to distract him from his pride and affection for her.

"Leah-Ann, do you see anything on the horizon that might challenge the future work of the Down to Earthers?"

She considered this. "I would say that the most serious challenge to our gains is coming from those damned militias in the third world. That is where our biggest challenge will be in keeping the momentum toward 90 percent emission cuts. Worse, if we cannot restructure their industrial development in a sustainable manner, the industrialization of the third world could throw us back into a planet subject to fossil-fuel emissions again."

Mitch and Leah-Ann continued their lunch largely in silence. It was not an uncomfortable silence; it was thoughtful and reflective.

Leah-Ann broke their reverie by tapping Mitch on the shoulder. "Well, you need to get ready for our first visit—we will be checking out the new power grid in the Buffalo, NY region. We will need you for your expertise in climate stabilization engineering. Get yourself packed and ready. On Friday, meet me at my Climate Center office at two p.m."

April 27, 2152

After packing for the journey to Buffalo, Mitch invited Gabriella to dinner. They went to a small cafe with a garden atmosphere.

He opened the conversation, signing, "You know, I really haven't had a chance to get to know you since you became my guide."

"It was sudden that I was called upon. Since I am a deaf-mute there is not much to get to know."

Mitch countered, "I suspect that there is a lot going on behind your sad golden eyes." She smiled sweetly but her sad eyes spoke along with her sign language.

"I am mostly in a silent world. People who can hear and speak generally keep me out of their conversations. So, I live alone in silence, though, to some extent, I can read language from the lips of people. However, most people do not face me to allow me to read their lips. So, I mostly live in my own head. It is not so bad. I can observe the world without chatter, but alone I am."

Mitch felt her loneliness. "I will try to share my conversations with you when I speak to others."

They continued a conversation through their dinner. Several times, Mitch drew a silent laugh from her. After they finished their meal, they walked to their rooms and said goodnight.

The next day they took a high-speed train to Buffalo, arriving around six p.m. In the morning, Mitch shared breakfast with Leah-Ann and Gabriella on the seventeenth floor of the Energy Monitoring Building in downtown Buffalo. It was a tall, thin, blue obelisk with arrays of sensors extending outward from each floor. He observed a stunning view of the urban landscape of Buffalo. The city was so futuristic. It appeared like something drawn out of EPCOT Center. The monorail system was spread out in a vast network throughout the city. Motor vehicles were absent. Fountains and gardens broke up the buildings and monorail tracks. The buildings were all new with the exception of the Old City Hall standing at the periphery of the urban landscape. The center of the city was dominated by a large steel and glass building whose widows were tinted with the super efficient color of the green, solar-absorbing glass that was their trademark. In the air above the building, the words "Energy Central, Buffalo, New York" flashed above the building. In the distance beyond the city, windmills dotted the landscape. Mitch signed. "This is nothing like the Buffalo where I worked in the early 2000s. There were windmills back then, but nothing like this."

Gabriella signed back to Mitch, "This is one of the cities that was mostly rebuilt from scratch."

Leah-Ann thought to herself, *I wish that I could understand sign language so I could get to know this lady a little better.*

Mitch looked at Leah-Ann, then at Gabriella. "When you speak, look at her. She can read lips." He then signed to his guide, "Gabriella, you can relay your conversations to Leah-Ann through me. This type of conversation may slow us down, but it will make us more of a threesome on these journeys."

The three of them went into the Energy Central building. Leah-Ann introduced Mitch and Gabriella to Dr. Jeanine Davis, a heavyset woman who carried the authority of a chief of operations. They took a seat in a small meeting room and shared the fruit elixirs that Mitch had begun to think of as a wonder drink. It was small talk in the beginning, but Mitch asked Dr. Davis to look directly at Gabriella when she spoke so that she could share in their conversations and told them that he would be a translator when Gabriella wanted to speak.

She smiled warmly at Mitch and signed, "That is so sweet of you to say that, but please tell Dr. Davis and Leah-Ann that it is not necessary as I will not ask questions or add conversations I am simply a guide for the journey." Mitch did so.

Dr. Davis spoke to Leah-Ann. "I suppose that you want to see our energy operations data from the past six months."

Mitch jumped into his active participant mode. "Actually, Dr. Davis, I am here to take a fresh look at your energy operations to see if we can further improve your efficiencies to approach 90 percent. I would like to visit the wind farms and the Niagara Falls power-generating center, but I would first like to see your carbon emissions data."

Leah-Ann spoke. "I will work with Mitch to explore how we can tweak your technology to improve the efficiencies of operation to meet the new 90 percent efficiency standard."

Dr. Davis shook her head, causing her wavy brown hair to swing rather abruptly. "Be my guest, but I really don't see how we can do better with our energy grid—the city and the towns beyond are already saturated with windmills."

Leah-Ann then responded, "Well, we'll see."

Davis grimaced and gave them directions to the data library.

When they arrived, Leah-Ann turned to Mitch and made sure that she was at eye level with Gabriella. "If Davis seems defensive, she is. Western New York is a prototype for the future of energy and transportation for other regions of the world. The transportation network was designed in the last century by a brilliant transportation engineer, Dr. Bertrand, who worked closely with a civil engineer, General Curtiss. Both participated on a technical panel in 2064 that set the course for the national grassroots network. Between 2065 and 2075, Curtiss assembled the best brains of the energy engineers of the world to create the design for this carbon-free city. The work of Buffalo and its surrounding region is a glowing success, but we can do better. The Chief of Operations in Buffalo carries a heavy responsibility. Davis does not see why it is necessary to ramp up the operations efficiencies from 66 percent to 90 percent, so she will be wary of our involvement."

Mitch turned to Gabriella and signed, "Did you get all of that?"

She fixed her gold irises on Mitch and signed, "Yes. But I want you to stop worrying about me. As your guide, I am meant to be in the background. Otherwise, I will just slow you down."

Mitch nodded and informed Leah-Ann of Gabriella's wishes.

He then turned to Gabriella and signed, "Are you aware of my interactions with Curtiss in the 2034 period of this dimension?"

Gabriella signed back, "Oh, yes. That is just before I took over as your guide when Reverend Tyndall was murdered. And I also know that you were very weary of working on committees. Perhaps you now can see how it laid the foundation for this brave new world. And look at what effect you had on Curtiss. By handling him with kid gloves and diplomacy, you turned him around to accept the grassroots movement and he became instrumental in building this energy infrastructure. Think of what may have happened if you had let him go down in humiliation following the defeat of his scheme in 2034."

Over the next ten days, Leah-Ann and Mitch labored in the library analyzing the energy production and consumption data. Leah-Ann then took Gabriella and Mitch to the sites of power generation.

The windmill farms were amazing. Miles and miles of a surreal setting of white windmills that now featured five small blades (that seemed to have a green veneer on them). Leah-Ann explained that the green veneer turned iridescent blue at night, repelling birds away from them. "During the day, the solar cell-infused blades provide the energy for them to turn, even when the wind isn't blowing." The windmills were mobile so that they could change their orientation to the coming winds. Leah-Ann commented on these gargantuan structures. "The Down to Earth coalition of western New York worked largely off of philanthropic donations to create this city as a model of energy and transportation."

Mitch looked at the landscape of almost 1,000 windmills. *It is amazing what they have accomplished with the government staying in the background.* A burning question flowed out of him. "Leah-Ann, hasn't there been an outcry against these windmill farms by the grassroots groups regarding bird mortalities?"

Leah-Ann smiled. "Remember, Mitch, these structures are a result of the Down to Earthers. In the 2050s, they were split regarding the need for more green energy versus the death of birds flying into the blades. They ordered a study of how windmills could be made more bird-friendly. The breakthrough came in 2075, when a new design that emits sounds that are out of the range of our hearing frequency but are heard—as an ear-piercing irritation—by birds. They also imbedded blue glowing disks in their blades, which were found to be a signal to avoid and divert their flight patterns around the farms. Lastly, when they installed the micro-focusing solar panels throughout the windmills, this had the side-benefit of keeping birds away by the blinding light of the reflecting sun off the mirrors. Now we are averaging ten to fifteen bird deaths per year, down from 4,000 in the 2050s."

Leah-Ann's team took measurements of air flow and passed the data over to her engineering construction team to retrofit twenty of the windmills with new blades that were a vastly different shape from the conventional shape that had lasted for hundreds of years, since the conventional windmill was invented. The owners of these windmill farms gladly bought into this upgrade, as they would get the new systems at a cost that was shared by the national grassroots network. The design, developed at Leah-Ann's Climate Center Lab in Albany, concentrated swirls of wind energy that had constant exposure to the head-on incidence of incoming wind into the turbines. It was expected that the new blades would increase their efficiencies by 10–15 percent.

The initial prototype project focused on the retrofit of the twenty windmills participating in the project. Leah-Ann directed her climate engineering team to change the blades and adjust the operation of the rotating windmills to interface with the needs of the channeling of wind flow. This pilot operation would determine how the system could be further tweaked and if it was justifiable—on a cost-versus-energy-efficiency basis, as well as a significant yield of greenhouse-gas reduction. If the numbers worked, it would justify a major retrofit of all 1,000 windmills here in the next five years.

Mitch, in the meantime, gathered data to see how many windmill retrofits would be needed worldwide to replace fossil-fuel energy and reduce greenhouse-gas emissions by an additional 10 percent. Every 5 percent increase in the efficiency and power generation by windmill farms nation wide would result in a retirement of one of the forty-five remaining fossil-fuel plants. However, the big impact would be in the third world and the developing world, where their new need for more power in their transition to modern society could start with more wind power and less of an incentive to go the old way of fossil fuels. They could do this analysis thanks to the precise monitoring records from the past hundred years of each windmill's efficiency.

Leah-Ann and Mitch were encouraged by their theoretical models and readied the twenty sites for operation. After a month of work, they justified the initiation of the pilot project. On the ceremonial "kick-off day," they were accompanied by Dr. Davis, who was still skeptical.

"I still think that this is all a waste of time and energy," she said.

Leah-Ann responded, "Wait for our final report and you will see the positive impact that this site will make on the worldwide consumption of green versus fossil-fuel energy production."

The following month, Leah-Ann took Mitch and Gabriella to the Niagara Falls hydroelectric facility, which now provided green energy to much of the northeast. They were evaluating what it would take to extend the grid to the Midwest, where coal plants had been largely retired over the past century. However, the region still relied on natural gas for power. As with the windmill evaluations, Leah-Ann looked at what it would take to expand the Robert Moses power plant, which would require a further diversion of the Niagara River above the Falls and expansion of the existing turbines. It had been previously upgraded in 2090 with the construction of a new diversion pipe from the falls. However, there was now a schism between the National Clearinghouse—that fought for every increase in the reduction of CO_2—and the local grassroots people who opposed any further diversion of the Niagara River for environmental reasons.

Leah explained to Mitch, "We are here to determine if a prototype project is worth it from both environmental and energy production perspectives. Ultimately, it would be the locals who would (or would not) buy into the project."

"Roger that," said Mitch. "Let's roll up our sleeves and get to work."

Leah-Ann smiled at this and led him into the field, where they did their analysis over the next three weeks.

In the end, Mitch and Leah-Ann determined that the incremental increase of energy power production was not significant enough to outweigh the environmental damage of robbing flow from the Niagara River and the Falls. Mitch had determined that, at best, the expansion of the Robert Moses facility may reduce carbon emissions by 0.1–0.2 percent with no apparent worldwide benefits. Hence, they would recommend that this project was not worth further analysis.

Their last stop was the western New York "Carbon Sequestration Project (CSP)." Leah explained. "The CSP is a fancy term for a tree farm that was created in the suburb of Lackawanna, on publicly owned land; 10,000 acres of industrial wastelands were planted with vegetation that would absorb CO_2 from the atmosphere.[64] It is considered to be a world model, and we have been monitoring the CO_2 uptake, and it is making a remarkable

---

64 Plants absorb CO_2 from the atmosphere through their stomates as an essential gas to drive their metabolism.

effect on lowering CO_2.[65] We are not here for an evaluation, just a field visit, as I am very proud at the success of a simple project that is understandable to the general public." At the end of the western New York visit, they prepared their report and presented it to Dr. Davis. She grumbled at the concluding recommendation to start the windmill pilot project.

"Well, you will do what you have to do. I'm still not convinced."

Mitch then focused on his section that showed the reduction in greenhouse gases worldwide, if other nations followed the Buffalo Wind Model. This seemed to invoke a change in Dr. Davis.

Davis smiled. "I never thought about it in that way: 'The Buffalo Wind Model.' I like that!"

Davis was also surprised about their recommendation for no action on expanding the Robert Moses hydroelectric plant. "Well, I have to say that this shows that you have been objective. I tip my hat to your work."

After they left Davis's office, Leah-Ann was pleased. "Good work, Grandpa. You turned her around with your term 'Buffalo Wind Model.' That was brilliant. She could have gummed up the works by inciting the local grassroots community against our recommendation. But now, we travel to the Midwest. Specifically, we journey to Nebraska, to view the agriculture initiative and to see if we can help with their new mycorrhiza[66] infusion."

June 3, 2152

The threesome traveled to Nebraska by a high-speed electric train. On the journey, they enjoyed each other's company and laughed a lot. It was good to leave their work assignment alone for a while. At one point, Gabriella had Mitch in stitches with her attempts to teach Leah-Ann sign language. At the end of the ride, Leah-Ann briefed Mitch on what his mission would be in working with individual farmers to offer them a no-cost participation for their pilot project aimed at increasing production and minimizing use of water. Mitch's eyes lit up. He knew that he was about to assume the role of a supporting manager of agronomic engineering.

"I can really get into that. I have been stuck indoors with too many panels, task forces, and forums during these journeys. Besides, my grandfather was a dirt farmer in the Midwest."

Leah-Ann smiled. "Great. I am glad to have you on board, Grandpa."

---

65   A recent article by the UN Food and Agricultural Organization entitled "*To what extent could planting trees help solve climate change*" describes how this can work. See https://www.theguardian.com/environment/2012/nov/29/planting-trees-climate-change
66   See https://en.wikipedia.org/wiki/Mycorrhiza

They arrived at Lincoln at two in the afternoon and were greeted by a woman and two men. The woman approached Leah-Ann.

"Dr. Stuart-Warner? I am Jenny Rosen, Chief of Operations for the Midwest Regional Farming Co-op. I don't know if you remember me from the agriculture ecology conference a few years ago. Anyway, I will be your host for your visit here."

Leah-Ann responded, "Hi, Jenny. Of course I remember! I enjoyed your presentation on the mechanics of establishing and operating a farm coop."

Introductions were made. Jenny was middle aged, with sparkling eyes and a smiling face. She was dressed in blue jeans and a blue and white polka-dot blouse. Jenny led the group to an electric bus that took them to the outskirts of Lincoln. Other than open fields and farmland, there was nothing on the flat horizon until they saw a red structure in the far distance. As they approached, the red structure turned out to be a barn—a giant barn, nothing like Mitch had ever seen.

"Wow, you must have a ton of cattle in there, Jenny," he quipped. Everyone laughed.

She explained, "The barn appearance is actually a facade—not my idea. Inside, it houses laboratories, offices, and presentation rooms."

Leah-Ann looked at Jenny. "So what is on our agenda today?"

"Well," Jenny responded, "I've been briefed on your mission to pilot processes to make our farming coop more efficient in production. I am all ears—particularly on the mycorrhiza[67] cultures. However, I must tell you, the farmers on our coop and other local farms are apprehensive. They have a mindset that the grassroots movement's transformation of agriculture to small farming units was all the success that we needed. They all follow the principles of sustainable farming. The new technology is something they cannot connect to—they view it as the National Down to Earth Center's attempt to be the federal government. You will have a challenge in turning them around to your proposal. As for myself—I am all for it and you should look at me as your interface. Though, I would advise you to work with them as opposed to preach to them."

Jenny looked at Mitch and Leah-Ann, who were absorbing what she said.

Leah-Ann smiled. "I appreciate your open-minded attitude . . . and yes, we do want to help as advisers, not imposers."

Jenny smiled, too. "Good. So, for tomorrow, I will be throwing you into the fire. I have invited all of the local farmers and representatives from the two remaining Big Ag industries in Nebraska. I am told that you are prepared to explain your proposed technological improvements. I thought that you should start with the full group and then you can break out to work individually with our farmers to establish your pilot projects."

---

67 See https://en.wikipedia.org/wiki/Mycorrhiza

Leah-Ann responded, "That is perfect. I will open with my virtual 3D empowerment presentation; then Mitch will explain the mechanics of how the newly enhanced mycorrhiza cultures work and the benefits of adding the cultures by drip irrigation process."

Jenny looked skeptical. "I hope that this will not be high-tech and slick. These folks want to remain as simple dirt farmers with their focus on green practices and gradual enhancements as the climate problem continues to cool down—and cooling is their expectation. One more item—you will need to keep your presentation to one hour."

Mitch and Leah-Ann looked at each other apprehensively. He responded. "Well, we will do our best. Our main hook will be good old-fashioned money. With Down to Earth funding, they can adopt our technologies and increase their production while saving precious water."

"So you want to buy them off? I can tell you—that is the way they will see it!" Jenny snapped.

Mitch turned crimson red and his cheek twitched. He looked at Gabriella, who signed to him, "Let it Be."

Further calming Mitch from an explosive retort was Leah-Ann, who intervened. "Jenny, I do not think that Mitch intended to convey the impression that we will *force* the technology on their farms. It will be provided to only those who want to participate in the pilot study."

Jenny sighed. "I guess that my response was defensive—defensive on behalf of our local farmers. Mitch, I'm sorry that I jumped down your throat. Why don't we have dinner and plan our presentation for tomorrow?"

And so they did. Mitch and Leah-Ann shortened their talks and Mitch substituted illustrations for words and simplified the technical aspects of mycorrhiza, explaining how the cultures are injected directly into the root zones of their crops.

The next morning, Granger Hall was filled with fifty people. Mitch knew, right away, that Jenny's admonition to simplify the explanations of the pilot project was spot on. *These are down-to-earth people in the true sense of the word*. Jenny had told them that they had their own simple creed: "We are fed by the earth, so we will treat the earth well." They were committed to sustainable farming, but the bottom line for them was still about profit: 'what will this do for my crop yields?'

Leah-Ann's presentation was short and sweet. She reinforced the national mood of commitment to the Down to Earth movement and summarized how it was spreading to other parts of the world. She made a motivational pitch about the contribution of the Down to Earth farming community.

She ended with the key statistic. "You—each and every one of you—has gone to bat for our climate. Collectively, our national emissions of greenhouse gases from agriculture have been reduced by 54 percent in the past 150 years. Now that may not seem like much compared to the U.S. overall reduction of 75 percent from all industrial and vehicular sources of greenhouse gases. But reducing the emissions from agriculture by using natural sources for fertilizers and using more efficient irrigation was always considered to be a hard nut to crack. We optimistically projected 20 percent reductions in 2050. So, you all have exceeded our expectations. But you can do better! Listen to Mitch's technical summary of the improvements that we are proposing. If you take this proposal seriously and allow us to take the first precious steps, we may be able to accelerate the reductions in farming emissions and catch up to the national performance standard of 85 percent reductions of nitrous oxide emissions, relative to 1990."

The audience applauded with at least a hint of enthusiasm. Yes, Leah-Ann had gotten them warmed up, but Mitch sensed skepticism about how their proposal would affect existing practices. Leah-Ann then introduced him. Dressed in blue jeans and a flannel shirt, he was ready and bounded up to the podium.

"Good morning, folks. How are you all doing?"

An undertone of disconcerting, mumbling words could be heard in the crowd along with whispers and shaking heads.

Mitch smiled slightly and lowered his head as the thought came to him: *That did not go over too well. I need to get 'down to earth!'*

He looked up, took a deep breath, and started again. "You all know that agriculture contributes a significant amount of greenhouse gases to our atmosphere. In 2014, it contributed 9 percent of the total from human activities in the U.S[68] and 18 percent worldwide. Today, agriculture emissions have been significantly reduced. Agriculture has been a difficult challenge. Yes, there has been progress, thanks to your grassroots effort at better manure management, more efficient irrigation processes, minimizing the application of fertilizers and pesticides, and smaller cropland areas as Big Ag is transitioning to small farms like yours here in Lincoln County. Even Big Ag has come around to adopt these practices. Farmers deserve a round of applause—you are truly a grassroots solution to the big picture of reducing greenhouse gas emissions worldwide."

There was a spontaneous applause from the audience.

Mitch applauded with them and then continued. "Yet, there is so much more that we can do, largely based on increasing the efficiency of producing crops that will further reduce emissions."

The audience suddenly became very quiet. Mitch read the dreaded skepticism in the air and paused and took a drink from his lemon elixir.

"I want to state up front that what we are proposing here is not mandatory. In fact, we are, like you, grassroots. We want you to see the possibilities before we proceed. However, it is, we think, a win-win idea. Our twin goals are to increase your crop yields and

---

68 See http://www3.epa.gov/climatechange/ghgemissions/sources/agriculture.html.

to further reduce the greenhouse gas emissions from agriculture, not just here in Lincoln County, but also across the nation and throughout the world. The first steps toward doing this could originate right here in your backyard."

Mitch looked around and sensed a spark of interest building. "So, let me get right to it. As you know, the big emission from agriculture is from nitrogen compounds from your fertilizers that escape to the atmosphere as nitrous oxide—a powerful greenhouse gas, much more so than CO_2. By adopting practices that have reduced the use of nitrogen fertilizers, you all have made a significant dent in reducing those gases through sustainable farming. What if I were to challenge you to totally eliminate nitrogen fertilizers?"

Rumbling from the audience told Mitch that he hit a sensitive nerve.

"What if I further told you that you would get better crop yields and lessen the costs of crop production while reducing your nitrogen emissions? Yes, it is possible. It is all about efficiencies in your farming practices. I want to be up front with you, on our motive. Our goal is to get you to take your good work in reducing nitrogen fertilizers one step further: eliminate them. Eliminate nitrogen fertilizers that emit nitrous oxide into the atmosphere. That is one of the pieces of solving our climate-change problem. We want to build up your toolbox. There are old technologies—such as drip irrigation and adding Mycorrhiza fungus cultures into the root zone to improve nitrogen uptake from the soils.[69] There are new technologies, such as building a sewage treatment plant for the county where all of the nutrient-rich treated sewage[70] will be recycled back to your farms for irrigation, rather than having it dumped into the rivers. There are the mega greenhouse complexes that will allow you to grow most of your crops in a football field-sized controlled environment that avoids the ravages of drought and allows you to continue farming in the winter. This is a win-win proposal: The earth and its atmosphere will win by eliminating chemical fertilizers, thus removing the discharges of nitrous oxides to the atmosphere and ammonia to the waterways. You would win by saving money in fertilizers and having better crop yields.

"Our proposal is simple. We will deploy a team of 'partners in small farm agriculture' to the county. The team will work with each of you individually to make the win-win farming improvements a reality to each of your farms. We will work collectively to share

---

69 Op. cit., EPA website see footnote 39.

70 Raw sewage is required to be treated to remove viruses and bacteria. The result is clean, nutrient rich water that is disposed in watercourses. Conceptually this use of treated recycled liquid sewage for fertilizing crops has been floated in the agricultural community. A pilot project has been created in Kansas City. See: http://harvestpublicmedia.org/article/flush-fertilizer-city-farms-recycle-waste-kansas-city There is, however, a powerful example of a project that uses recycled sewage for an environmentally beneficial use: The Wakodahatchee Wetlands in Florida. In 1996, The Palm Beach County Public Utilities Department used treated sewage to create a wetland. The outcome was a lush ecosystem of ponds and wet meadow that now support a plethora of birdlife. In our story, a community wide sewage recycling for agricultural fertilization is visualized. See http://www.pbcgov.com/waterutilities/wakodahatchee/what_is_wakodahatchee.htm

ideas and consider building the new engineering of recycling sewage treatment plants—RSTPs—or a series of package RSTPs—to irrigate and fertilize your crops. Again, none of this is mandatory.

"You have all heard of the twentieth-century concept of inoculating plants with mycorrhiza[71]—the fungus attached to roots that stores carbon. It also efficiently regulates nitrogen uptake and use for the plant plus it has the potential for decreasing plant disease. In the 2020s, it was largely concluded that it was not cost effective for most local farms. However, we have created a new evaluation and application technology that may work for your farms depending on your crops and your soil conditions. If the conditions of your farm are suitable, the inoculation of Mycorrhiza cultures in the root zones is now very simple and inexpensive. Similarly, drip irrigation has been around for a long time, but improvements in technology over the past twenty years allow a more cost-efficient installation and operation process."

Mitch looked at the audience and saw expressions of bewilderment or wonder, he couldn't tell which. *Well, this is better than seeing disgruntled faces.*

He continued, "This will significantly reduce the need for conventional irrigation and enrich your crops with a nutrient soup that will drip directly into the root zone."

By this time, the farmers in the audience were shouting questions and hands were up. Mitch had been hoping for this.

"If you could bear with me and hold your questions for a few minutes, I am almost through. I want you to see the full picture before I open it up for discussion."

The audience quieted down.

"As Leah-Ann and I both said from the start, none of this will be forced on you. We want to work with each and every one of you to share our toolbox of innovative technologies and common sense practices to improve your crop yield and save you production expenses. We will also install a network of micro-atmospheric monitoring stations for nitrous oxide. In the past, such incentives—such as the nitrogen stewardship programs earlier in the century[72]—were from the top. They were good programs at the time, with the federal partners offering farmers incentives to manage their nitrogen releases. But times have changed. We are the grassroots folks—just like you—and are not forcing any particular approach down your throats. We will be working as dirt farmers alongside you, not as bureaucrats. We simply want to work with you as fellow farmers to see what might be done to improve the productivity of your crops, improve the efficiency of irrigation, and vastly reduce or eliminate the use of nitrogen fertilizers."

A woman in the first row shouted out, "What is this going to cost me?"

---

71 See http://www.ars.usda.gov/SP2UserFiles/person/4947/Presentations/ScagelMycorrhizaeOhioCents06.pdf.
72 See http://archive.epa.gov/partners/web/html/, with reference to the Ruminant Livestock efficiency program.

Mitch had fully expected this to be the first question. "It will cost you nothing but your time in working with us during the pilot stage of the project. We will not propose any changes to your farming practices that will not have a reasonable payback period. We will only be advisors. When the pilot project concludes, the ultimate decision will be up to you to balance investment costs with the payoffs in the savings on water irrigation and fertilizers, and increase in crop yields. If you choose not to participate, there will be no cost to you and you get free advice on your farming practices"

A man in a suit—the only one in the audience in business attire—had the next question. "Well this all sounds ideal, but what about the high-cost solutions? Building giant greenhouses and a centralized sewage treatment system will be expenses that trickle down to each of us, through the taxes of our local government."

Mitch was happy this question had come up. "Again, none of this will be forced down your throats. Collective decisions will be made in this room by all of you. No worries of penalties for not proceeding forward on our recommendations. The bottom line is the cost vs. benefit analysis that each of you make in evaluating whether these improvements will have a good payback period. Remember, we have something in common with all of you: we are grassroots people. We are one of you."

A woman dressed in a lab coat asked the next question, "I am Dr. Melanie Oritz and I am a an agricultural ecologist. What you said about 'micro-monitoring' makes no sense. You should know that you can not measure nitrous oxide or any gaseous emission from a regional source area. It can only be done worldwide, such as the ambient air measurements from Mount Mauna Kea CO_2 monitoring station in Hawaii.

Mitch was pleased that this question was asked and eagerly explained how it worked. "Agreed, Dr. Oritz. That cannot be done. But when I say "micro-monitoring,' it involves test plots where the entire crop area will be enclosed by a chamber where emissions of nitrous oxide from the soils and plants can be measured, before and after the practices are implemented.[73]

The questions went on for an hour and a half and were deftly handled by Mitch and Leah-Ann. At the end of the session, they were exhausted. However, they were encouraged by the interest and active discussion. By this time, eight families had already applied for the pilot program. For the next two weeks, Mitch worked tirelessly, speaking to farm owners, taking soil tests and working with engineers to examine the farms of the applicants who volunteered for the pilot-testing program. Every day was a new adventure. At night, the threesome—Mitch, Gabriella, and Leah-Ann—had dinner with Amos and

---

[73] In 1977, the Author conducted such a "micro-measurement" project, enclosing the common reed, *Phragmites sp* in a glass enclosure and collecting air samples, measuring mercury emissions from the enclosed airspace. See: http://www.nature.com/nature/journal/v274/n5670/abs/274468a0.html
Based upon these emissions from one plant, he further went on to estimate the emissions of mercury over a larger area around Onondaga Lake in Syracuse, N.Y.

Andrea Hutchings, who were putting them up in a vacant cottage behind their farmhouse. They were one of the families who would be participating in the study. They sometimes discussed the details of what they would be doing over the next year, but mostly the five of them just had fun: playing card games, taking walks at dusk, sitting on the Hutchings' glider on their porch into the evening, and laughing a lot. Leah-Ann was getting the hang of sign language and her antics with Gabriella always brought a smile if not a laugh. There were day-to-day problems, but Mitch was fully engaged and loving it.

Mitch also made friends in the community. There was "Shoe"—sometimes referred to as "Old Shoe"—who was an artist and drew portraits of the scenes surrounding the pilot projects. At night, Shoe would entertain the locals with his guitar. Then there was Diana, a dancer who sponsored "community hoe-downs." Yes there was square dancing, but she also performed ballet to recordings of symphonic music. "Mr. G." was a designer and builder of log cabins. No one knew his real name. Whenever introduced, he identified himself as "G. Just call me G." Finally, there was Zeke, the community's noted fisherman. He had an uncanny show that he put on for the people in the area. Folks said that Old Zeke could smell a striped bass coming down the Platte River. And he showed them. He would stand by a bridge and patiently wait until the right moment when he would dive in the river to catch one in his arms—more often than not, he succeeded.

Then, one evening, Leah-Ann spoke privately to Mitch and Gabriella.

"Mitch, I know that you have enjoyed getting your hands dirty in this creative project, but it is time for us to move on."

Mitch had expected this, of course, but was disappointed that this could not be his permanent career here. On the other hand, he needed to stick with the program, since he felt a pull back to 2020, missing his family. So, the threesome said their good-byes to Amos and Andrea and flew to Nigeria. As he waved good-bye to them from their vehicle, he was melancholy about leaving them and the whole project in Nebraska.

September 15, 2152

They landed in Abuja, Nigeria and took a bus to the Borno region. They were greeted by three very tall Africans who introduced themselves as Tuo, Patsie, and Imbibe. They all spoke English. Imbibe had a distinctive British accent that seemed strange. They told the threesome that they would be their guides to the troubled area.

They all jumped into a small bus and traveled eastward. On the bus, Gabriella discovered that Tuo knew sign language and she enjoyed a signing conversation with him during the entire trip, periodically breaking Tuo into giggling fits. Mitch, however, was serious and silently thoughtful.

Finally, he came out of his reverie and asked, "Is someone going to tell me what this trouble is all about?"

Imbibe filled him in. "Well, no doubt you know of the loss of a very large tract of our rainforest over the past several months. This was caused by the Otubo—the rebel group who is attacking our progressive projects. We seek to maintain ecological integrity and to develop the legume-harvesting practices. The Otubos just want to protect their monopoly over the old unsustainable practices of clear cutting, farming without erosion control and over-fertilization without regard to the environment."

Patsie broke into the conversation. "The Kobe rainforests are native homes to natural legumes. The legume harvest project has gained worldwide attention since it taps the cultivation in an ecologically friendly manner."[74]

Tuo interrupted his signing party with Gabriella and jumped into the conversation. "The legumes play an important role in fixing nitrogen out of the atmosphere and delivering it to the crop. All of this was good in absorbing nitrous oxides out of the atmosphere. On the other hand, the burning down of rain forests released large amounts of CO_2."

Patsie raised his hand in authority. "You must be wondering about our forest loss. Many, many acres of our rainforest—including 60 percent of the areas cultivated with legumes—have burned down from the Otubo raids in August. We are sad about this, but our national conservation troops have combed the region, made many arrests, and battled with the Otubos, who now seem to have slipped into the interior of the rainforest."

Imbibe looked at the threesome with pleading eyes. "We so much need your help in working with our people to restore the legume industry. Our own 'grassroots' people—'Down to Earth,' as you call it in America—were very dependent upon the legume crop, since the legume agriculture industry is slowly bringing our nation into the twenty-second-century economy."

Patsie spoke. "Enough for now. We will discuss this further with our conservation guides tonight."

Mitch spoke up. "But you must tell us how safe we will be. Are the Otubos still raiding the area?"

Patsie responded, "They are a reactionary militant tribe, resisting the change of old ways. We are in perpetual war with these shegiyas – pardon my profanity They have been a fringe group of bandits and can attack at any time Yet, I believe that they have now scattered to the winds after our military sweep of the area. No worries. You will be accompanied by a squad from our elite nationals—an army second to none. The Otubos travel in bands of twenty to thirty. We will have with us one hundred nationals."

---

74 See Biology and Fertility of Soils, Z. Zhong, L. Nelson, R.Lemke; Biology and Fertility of Soils; August 2011, Volume 47, Issue 6, pp 687-699

Mitch looked skeptical. He had an uncomfortable premonition about this venture into the rainforest dens of the Otubo guerillas.

Gabriella signed to Tuo, "What is shegiya?"

Tuo signed back, "I keep quiet on that. I do not want to offend the lady with our dirty words."

By late afternoon, they arrived in the Borno region and stopped at Maiduguri, its capital. That evening, they shared dinner with the local Conservation Corps leaders and were briefed on the challenge of keeping the Otubo rebels in check and re-establishing the rainforest and the legume plots within the forest. The threesome would be taken to the heart of the devastated forest tomorrow, along with Imbibe, Tuo, and Patsie. They would be accompanied by one hundred soldiers and their guides.

That evening, they had a period of levity and laughter. Imbibe was his joking self and stole the show. Mitch was distracted. He noticed Gabriella signing to Tuo about a harmonica. Mitch was enjoying his camaraderie with Leah-Ann and Gabriella, but he had a bad feeling about taking them into a zone that was in a state of guerrilla warfare. Although he had been reassured by the Conservation Corps leaders that the Otubos had been "driven back into their holes," he still had an unsettled feeling about tomorrow.

His thoughts were interrupted by music. Imbibe was playing a guitar while Tuo was pounding a drum. Patsie started singing a native folk tune. Gabriella soon joined in by playing a mean harmonica. She was really quite good. Leah-Ann turned to Mitch and asked how she could play without the ability to hear her own music. Mitch shrugged. My sister was a deaf mute and had the ability to play the piano by sensing the vibrations of the music she played on the floor below her.[75]

Leah-Ann's question was answered by Artu, one of the Conservation Corps leaders. "I know this. We had a speechless and no-hear person in our corps who had learned to play flute. She has an implanted device in her right temple that allows her to amplify her sense vibrations. She can actually hear many sounds. I suspect that it is same with Gabriella."

Amid the joviality of the evening by the campfire, Mitch momentarily lost all of the worries that had plagued him and just enjoyed the camaraderie, particularly with Gabriella and Leah-Ann.

The next day, they started at the crack of dawn and traveled to the destroyed rainforest. On the way, Leah-Ann, Gabriella, and Mitch, along with the Conservation Corps stopped at one of the non-burned forest areas and showed them a local plot of legumes in an area that had been spared from the burn-downs. It was a marvel. The legumes were integrated within the forest, which was characterized by interspersed meadows that supported the growth of cowpeas and other legumes. The landscape was a stunning maze of

---

75 When he lost his hearing Beethoven continued to play and compose by sensing the vibrations of the piano on the floor

forest, savannas, and streams. The legume plots were intertwined into the meadows but did not overwhelm their habitats, as in the past.

Imbibe explained the farming system. "When the local conservation movement took root at the end of the last century, it was met with much resistance. The pilot studies that were guided by world conservationists from your grassroots groups demonstrated that in the long run, it was better to live with nature rather than dominate it. Over the past hundred years, there has been a very steady sustainable yield without destroying the habitat by over-exploitation. Still, there was local resistance that eventually became organized resistance under the group that became known as the Otubos. The Otubos want to return to old slash-and-burn habits that are now against the law to maximize their crop production without regard to the environment.[76]

They returned to the bus and drove on toward the burned area. Normally, Mitch accepted and appreciated forest fires as nature's way of recycling itself. However, this purposeful burn brought back the prospect of the world returning to the century-old practices that had destroyed vast areas of habitat and contributed to warming of the climate. They walked through the burn zone and then to a location of a rocky wall, which had an uphill pathway. Twenty of the hundred federal army accompanied them into the rainforest while eighty held back, fanning out in all areas to scout the area for Otuba militants.

Patsie instructed the group, "We go this way, follow me. At the top of this hill you will see a panoramic view of the forest and the burned zone."

He started the climb followed by the rest of the group. Suddenly, the air was filled with the sound of rifle shots. This was most certainly an Otubo attack. Patsie went down with a bullet that hit him square in the back. Mitch thought he saw Imbibe go down as well. The army escort returned fire and the lead guard ordered Arturo to take the Conservation Corps and their three guests to safety in the caves. Mitch shuddered as a bullet shattered the rocks just inches away from his head. He, Leah-Ann, and Gabriella continued following Arturo along the base of the rocks, but gunfire was all around them. Arturo tugged at him and pointed to what appeared to be an enclosed area in a rock overhang and led Mitch, Gabriella, and Leah-Ann there. Suddenly, three pineapple-shaped objects landed in their sheltered cove. In a split second, with an infusion of terror-driven adrenaline, Gabriella lifted up Mitch and literally threw him out of their cave, then she jumped on Leah-Ann, sheltering her body, just as the grenades exploded;

Oh my God, no! Mitch ran into the cave opening.

He turned Gabriella over and she was clearly dead. The khaki outfit that she was wearing was transformed, in Mitch's mind, to the white flowing gown that she had when he first met her – but it was covered with blood. Leah-Ann, however, was breathing shallowly and bleeding from multiple wounds on her arms and legs. The worst wound was in

---

76 See http://www.ecologic.org/actions-issues/challenges/slash-burn-agriculture/.

her side that was oozing blood. She opened her eyes and whispered, "I know that you are my great grand ancestor." Then she closed her eyes and slipped into a coma. Blood was pouring profusely from the place on her hand where her fingers used to be. Although her eyes were open, she was not there. She was shivering in her comatose state. As the chaos raged on, Mitch tended to her for the next two hours, wrapping her hand tightly in a handkerchief to stop the bleeding, applying pressure to the bleeding wound in her side and covering her with his own body when the debris from the shells came close.

In a twenty-minute period that seemed like an eternity, the rest of the one hundred army guards finally arrived and took over. Fortunately, there was a medic in the crew who treated the wound (injecting her with a localized medication that immediately curtailed the bleed-out) and supervised the transport of Leah-Ann to the helicopter. Mitch, despite his fear of flying in helicopters, jumped in. They were flown to a hospital in Abujo, where she was rushed through the emergency room into immediate surgery.

Mitch waited solemnly for word of her condition. As the shock and denial wore off, he suddenly dealt with the reality that Gabriella was gone and Leah-Ann was, most likely, dying. In the midst of his fugue state, he felt someone stroking his arm. A beautiful black woman was trying to comfort him. When he first noticed her eyes, his reaction was that they were kind. His second reaction was shock. The irises of her eyes were iridescent gold, like Gabriella's.

The lovely black woman put her hand on Mitch's face and quietly whispered, "Let it Be."

That opened the floodgate of tears. He sobbed and sobbed for what seemed like hours, with the woman cradling his head. He realized that he was on the floor.

He heard her whisper, "Look, look. Doctor is trying to speak."

Mitch looked up, then stood and faced the surgeon. He was a tall and heavyset African man with gray eyes. "I understand that Miss Leah-Ann was your associate and I am very sorry." Mitch braced himself for the inevitable news.

"Miss Leah-Ann is in a coma. She is out of danger from bleeding out, but she has a concussion and multiple shocks to her body. It will be very touch and go on her survival."

Mitch vowed to stay in Abujo on a daily vigil to tend for his great-times-six granddaughter. Finally, she was released from intensive care. He visited Leah-Ann for hours every day, until he was instructed to leave. When he was asked of his relationship to Leah-Ann, he simply said, "She is my granddaughter." The nurse looked bewildered by this improbable revelation, considering that Mitch's youthful thirty-something did not match the timeframe of being the grandfather of Leah-Ann.

It was a sunny afternoon. The days turned to weeks and after one month he thought that he heard something. Mitch had been sitting with Leah-Ann for an hour or so and was dozing. When he opened his eyes he saw Lea-Ann sitting up.

"Sleeping again, Grandpa?" was all he needed to jolt him out of his nap.

Mitch touched her face and felt like pinching himself to see if she were indeed real and that he wasn't dreaming. He called the nurses' station and they came running in. When the doctor came in and took Leah-Ann's vitals, he turned to Mitch and said, "I think that she's going to make it." Mitch spontaneously hugged the large doctor and then kissed Leah-Ann on the forehead.

Nine months later, after Leah-Ann's discharge and rehabilitation she returned to her office in Albany, where Mitch was managing the operations in her absence. Mitch was briefing her on what had transpired while she was in recovery.

Finally, she said: "Well, that was some ordeal we went through, but I am chomping at the bit to get back to work."

Mitch nodded, "I appreciate your eagerness, but a little bit at a time. The doc has advised that we start you on the way back to your work life by simple briefings on what you have missed. So, I am happy to report to you that the nationals have now taken full control of the rain forest and the local citizens have made an impressive start to reforestation of the burned out areas. Our Buffalo project has taken root and the windmills look like they are significantly improving their energy capture efficiencies. And the conservation teams are working with the farmers in the pilot projects in Nebraska. The enhanced mycorrhiza nitrification process however, has hit a stumbling block. They are having trouble adapting the cultures to enrich the soil micro-zones around the crop roots."

Leah-Ann raised the hand that was destroyed by the grenade and amazingly, it miraculously had new fingers on it.[77] Leah-Ann noticed Mitch staring at her hand when he gently reached for it as if it was a delicate butterfly.

"It's okay Mitch. Haven't you ever seen DNA stem cell retrogression before? Oh wait, you may not have."

Mitch looked up with soft eyes and said, "Well yes I had heard about it, but this . . ."

Leah-Ann chuckled, "Mitch, I know that you are from a different dimension, whatever that means, so let me fill you in. It's amazing isn't it?"

Mitch interrupted, "Leah-Ann, how did they get around the DNA sequencing for limb loss problems?"

"Well Mitch, you will like this. From the initial stem-cell concept they were able to take multiple minute slices of existing cellular tissue from the missing appendage and re-sequence the age specific genetics to call up, so to speak, the cellular memory and grow only that specific missing appendage."

"Wow." Mitch was silent and just kept looking at her hand. Leah-Ann gently pulled her hand away from his and put it on his shoulder.

---

77 There is current research on how to regenerate damaged heart tissue. See http://stemcells.nih.gov/info/basics/pages/basics6.aspx in 100 years, it may be possible to regenerate other damaged tissues, such as a severed hand.

Mitch thought, *Leah-Ann's skin reminds me of Leah's skin.* His memory of Leah was still very much intact.

"Well, Grandpa, this is as good a time as any to discuss you and your future in our world."

Mitch again wondered what context she was using the word "grandpa." He had not discussed her words that she had revealed to him at the moment of her near death. He assumed that she had not remembered and that this knowledge of her recognition of him as her ancestor is something that may be buried in her subconscious. But who knew? He decided to Let it Be. He was yanked out of his reverie by Leah-Ann calling him.

"Mitch! Mitch! Earth to Grandpa."

Mitch looked up at her. "I thought that I lost you in whatever space that you drifted into. I was talking about you. I can use you on this project. I want you to give some thought about being assigned to the western New York project, the Nebraska Agriculture enhancement project, or right here with me at the climate center, or Nigeria for that matter."

Mitch looked at her for a long while, "Well, I certainly rule out the Nigeria assignment. Being back in that jungle would bring back a PTSD of those awful last moments."

Leah-Ann looked at Mitch with sympathetic eyes. "I know that it must be rough on you. Gabriella was a gem. The three of us really hit it off and I was starting to envision her as your assistant if I could recruit you. But we must move on. I do want to recruit you." Leah-Ann reflected in silence for a while.

"Mitch, I suspect your melancholy goes beyond the loss of Gabriella. I sense that you have lost something from your past. I won't probe, but if that is so, it is still a good idea to look to your future."

Mitch looked at Leah-Ann and thought to himself that his descendant *was* his future in more ways than one.

"Let me think about this for a few days, Leah-Ann."

She responded, "Take all the time you need. All three assignments will wait for you to choose."

Mitch spent the next few days reflecting on what he had gained and what he had given, on journeys through the seven doors, and what he had lost. He still clung to the hope that he would be greeted by a new guide who would take him back to his 'dimension' of 2020. He missed his wife, Leah, and his daughter, Susannah deeply. Yet it was becoming increasingly clear to him that he was to be stuck in this time and dimension of the seventh door. It could be worse. He was truly inspired by this grassroots world and very anxious to get his hands dirty in prodding the Down to Earth movement forward to achieve even greater reductions of greenhouse gases. The people in my dimension were stuck in their complacency, denial and hopelessness. Mitch thought to himself, *The folks of this seventh door are moving forward, because the change is real to them. They are moving forward on their own, right at the grassroots.*

Mitch sighed and said out loud. "Yes, it is time for me to decide."

October 4, 2152

In the open farm fields of Nebraska, Mitch took a moment to enjoy the feeling of being outdoors all day. It had been two weeks since he told Leah-Ann of his decision to take on the farm project. Yes, she had been disappointed and surprised, since his background was in climate change monitoring. But he sensed that she knew that they needed space. He could still visit her and she had already come to Nebraska to visit him. But the weight of her presence as his great-times-six granddaughter would be too much for him and a constant reminder of the family that he left behind in 2020.

And then there were his hands. Yes his hands were dirty, from testing soils, injecting different variants of mycorrhiza cultures into the root zones and his examination of roots under the microscope at the lab. But he was getting his hands dirty in yet another way: his lifestyle. He was getting down to earth.

In the distance he could see someone—no it was a group of people—walking down the long straight road on the horizon toward him. Mitch rubbed his eyes.

Where in the hell were they coming from? There isn't another farm for ten miles in that direction. They must have had a breakdown in whatever vehicle in which they were traveling.

Mitch started walking down the road toward them and saw that there were seven of them. A woman and an adolescent girl were walking. The rest of them seemed suspended above them. As they approached, five of them started to dissolve into thin air, one by one. Could it be? Mitch strained his eyes and was reasonably sure that the vanishing people were Casey, Jake, Janey, Reverend Tyndall, and Gabriella . . . perhaps. After all these journeys, he could not yet discount his imagination playing tricks on him. Yet, as they were disappearing, the five guides suspended above the two women were waving at him.

Mitch wondered, Why couldn't they stay in the picture and guide me back to his 2020 dimension?

Then he focused on the two approaching women who were smiling and then running toward him.

"My God, could this really be happening?" Mitch exclaimed aloud.

Then the Leah of his 2020 world and his daughter Susannah embraced Mitch. He barely held control of his emotions as he hugged his 2020 family. He then gave in to tears of joy.

"You are here. You are really here!" said Mitch.

Leah responded, "Do you think that we would allow you to stay in this 'dimension' as you have been calling it, all by yourself?"

Susannah was holding Mitch tightly. "We are here to stay, Dad."

Following this tearful reunion, Mitch took them to the farmer's home and introduced them to Amos and his wife Andrea. "Time for an elixir," said Andrea. Mitch looked at the curious expressions on the faces of Leah and Susannah.

"Wait until you try this drink. It will knock your socks off and do not worry; it is natural and organic . . . like a tea," said Mitch.

After a long moment of sheer joy of their reunion, Mitch asked his wife and daughter, "So how did you get here?"

Leah explained. "Not quite sure, but we were suddenly on your trail throughout this journey. We were mostly following signs of you, but we periodically caught up and observed you from a cloud above, though you could not see us."

Mitch nodded, thinking about how he was suspended above Leah-Ann in the 2152 realities of other doors. "I know what you are talking about."

"Yeah, Daddy, once we were in a bubble above your bubble seeing you observe Leah-Ann and her family during Thanksgiving."

Mitch nodded. "But who guided you here?"

Leah explained: "At first it was just Casey, who you talked about incessantly after your first night's journey. Then, we were joined by the twins. The priest joined us after he was murdered in the 2052 of the seventh door. Finally, the sweet deaf-mute came on board at the end, after her demise in Nigeria."

Susannah joined in: "It was weird, Dad. They never spoke to us, but we knew they were there. It was as if they were moving us forward."

Leah finished the explanation: "I can best describe their presence as a 'stream of consciousness.' "

So, Mitch and his family settled in and accepted the gracious offer of Amos and Andrea to live in their cottage on their farm. Mitch was really into it—taking soil tests, developing the enhanced Mycorrhiza cultures, overseeing the evaluation of small community package sewage treatment plants that would recycle treated sewage to a regional drip irrigation system in their crop land, and working with the local farmers associated with the Down to Earth participants for the pilot studies. Leah was thrilled with involving herself with the project as Mitch's "right-hand lady" as he referred to her. Susannah, however, was unhappy with her school. She preferred an urban social setting. One evening Mitch and Leah discussed the situation.

"We have really found our niche here but I worry about Susannah. She is so maladapted to living in Nebraska."

Mitch nodded. "You know, I have been thinking about this long and hard, and I may have a solution. What do you think about having her spend her senior year in Albany?"

Leah shook her head.

"You forget. Susannah is from 2020. Other students her age are on a different future wavelength. She is missing 130 years of a future world. It works here in Nebraska for some reason, but I fear that she will be lost in a city."

Mitch smiled. "Do you notice how intuitively you have fit into this world, even though we are on a farm?"

Leah thought about this. "You know, when I think about it, I feel like I became fully immersed in this culture, almost as if I have carried 130 years of knowledge with me."

Mitch clarified, "That is because you have. You are in the participant role, and you are up-to-date. So is Susannah."

After a thoughtful silence, Mitch asked, "So what do you think?"

Leah smiled. "Susannah would certainly like that—she was charmed by Albany when we followed you there—but are you really willing to give up your work on this project?"

Mitch explained his plan. "We do not have to do that. I was thinking about talking to our great-times-six granddaughter and seeing if I can arrange a modification of my assignment. I thought that doing a split assignment—spending six months here, alternating with six months at her climate-monitoring center in Albany might be a win-win situation. As much as I love getting my hands dirty in the project here, there are times when I miss the technical aspects of having my finger on the pulse on the global picture of our efforts to maintain the recovery from the mess that we had made with our climate in 2020."

Leah looked skeptical. "Well first of all, where would Susannah stay while we are in our periods in Nebraska? And would Leah-Ann go for this change in your venue?"

Mitch smiled. "Same answer to both questions. Leah-Ann will certainly accept this scheme—she has given me full independence in defining how I spend my time with the Down to Earth Movement. And I believe that she would put up Susannah in her own 2162 home while we are back here. She feels a bond with our family. I know that she suspects that we are her ancestors."

Leah smiled. "So, let's do it."

As Mitch suspected, Leah-Ann fully accepted the proposal for the modification of his assignment. In fact, she welcomed it with glee. When Mitch brought up about lodging Susannah while he was back on the farm, she didn't hesitate.

"Of course, Susannah can stay with our family. We will love her as if she were our child."

Mitch and Leah took a long pause and Mitch guessed what his descendant, Leah-Ann was thinking—the same as he was thinking: *As if she were your own great-times-five grandmother.*

Later, Mitch and Leah were walking at the edge of Amos and Andrea's farm field. Mitch felt sad thinking about Gabriella. Leah looked at him, sensed his melancholy and knew why he was sad.

"I know." she said. "Susannah and I crossed the path of your period with Gabriella. We both knew that there was something very profound and spiritual in her presence on this journey. But, Mitch, close your eyes for a minute." Mitch did so and felt a serene presence and his sadness dissolved. Leah smiled at Mitch. "Well?"

Mitch took a moment to respond. "It felt like Gabriella was standing right in front of me."

Leah stroked Mitch's long hair. "That is all you need to know."

On their final night on Amos and Andrea's farm, the family sat on the swing glider in the pleasant mild summer evening under a star-lit sky.

Susannah said, "I will miss this place. It is so quaintly rural and I loved staying with our hosts in their back cottage."

Mitch replied. "Well, we will be back here in just six short months. Besides, you wouldn't want to spend your winter in Nebraska."

Leah asked her daughter, "Susannah, do you ever miss being back in 2020?"

Susannah paused to reflection. "You know, Mom, I don't. And I strangely suspect that there is some part of our family that is still there."

After another pause amid the chorus of the crickets, Mitch thought, *So, we Let it Be*. Mitch thought about the lyrics of the now-familiar song that had been playing in his head during all of his journeys.

Endnotes, Chapter 7:

1. Local grassroots sustainability groups are starting to grow into a movement throughout the U.S.[78] The local grassroots sustainability movement is also growing worldwide.
2. As an example of one of these grassroots coalitions, the village of Bedford, NY is a coalition of citizens who are reaching out to their community. The mission of the Bedford 2020 coalition is to lead, organize, and promote a community-wide effort to reduce greenhouse gas emissions 20 percent by 2020 and to create a sustainable community that conserves its natural resources. They have engaged the surrounding community in projects that include incentivizing solar energy use, home energy efficiency retrofits, waste reduction and increased recycling, education around electric vehicles, and local food projects.[79]

---

78 See http://www.sustainablemeasures.com/projects/Sus/Sustainability/5
79 See http://bedford2020.org/

3. Sustainable Westchester[80] is another grassroots group. They have recently arranged to provide energy to the citizens of sixteen municipalities through a community-based bulk-energy purchasing program intended to lower costs and increase the use of renewable energy in Westchester County, New York.
4. Sustainable farming[81] is a growing movement throughout the nation. It encourages water conservation and minimal use of fertilizers. In addition to combating drought, less use of water on farms will involve a lower amount of energy needed to produce crops, which will translate to less greenhouse gas emissions, since less water will need to be produced from utilities, lowering their greenhouse gas footprint.
5. The minimization of fertilizer use will, as our story illustrates, result in lower emissions of a powerful greenhouse gas (more powerful than CO_2): nitrous oxide.[82]
6. Western New York, with westerly winds blowing off Lake Erie, is indeed a prime location for windmills and a drive to Buffalo from Route 20A reveals a landscape of giant windmills already in place. Buffalo's shoreline on Lake Erie is also a landscape of windmills. Wind power is growing throughout the U.S. Another example is the new offshore wind farms that is powering up off Block Island in Rhode Island.[83]
7. It can be argued that the entire scenario of increasing emissions reduction from 80% to 90 or 100% is unrealistic. However, this utopian effort described in Door # 7, is meant to portray the momentum that is needed in the worldwide effort to continue reducing greenhouse gas emissions lower and lower.
8. In December 2015, the nations of the world approved a climate accord with specific targets for each nation. That is, of course, top-down, but it is also non-binding. This is considered to be a weakness of the treaty, but it also puts the ball in the court of the grassroots. It is the people of each nation who will make the treaty work . . . or not.

    An often-overlooked aspect of the coming treaty is the aid package to developing nations. When implemented, it will send teams of specialists from the developed world that will assist in guiding the development of the third world in sustainable industrialization, commercialization, transportation, and residential living.

---

80 See http://sustainablewestchester.org/
81 See https://en.wikipedia.org/wiki/Sustainable_agriculture
82 See http:/environment.nationalgeographic.com/environment/habitats/sustainable-agriculture/.
83 See http://www.newsmax.com/Newsfront/wind-farm-rhode-island/2016/08/28/id/745515

EPILOGUE

Mitch awoke and knew that he was back in his 2020 bed. He thought to himself. *In my gut, I'd expected that I would end up here. However, I am not discombobulated. I am not disturbed, and I have no confusion, as I did after my return from previous journeys. The answers to the looming questions that would have thrown me for a loop after the first two nights of discovery are intuitively in my head:*

- **Was all of this just a dream?** Of course not.
- **Have I exited 2152?** No, a different part of me—or perhaps a different version of me—is still there, working with the farmers in the summer and with my descendant Leah-Ann in the winter. And yes, I believe that Leah and Susannah (or a different version of them) are with me in 2152.
- **Which version of me is the real Mitch?** The Mitch sitting in this bed is real, and so is the future Mitch. They just exist in different dimensions or "parallel universes."
- **What is the meaning of all of this and what is my mission now that all of these journeys are complete?** We need to take action, but with enlightened and open minds.

Mitch sensed the end to the stream of consciousness that has been the oracle of these journeys.

****

Mitch and Leah woke up at the same time. They looked at each other and started laughing. Mitch scratched his gray beard.

"I think that we both needed that to release the tension. Somehow, I know that you experienced the same dream."

Leah was still laughing, but tears ran down her cheek. "That was so intense . . . particularly the part where I doubted if I would see you again. But no, I do not think it was the same dream, except for the ending. Unlike the last journey, in this dream I was not observing you, but following you."

"There is one question that I have for you, Leah. At the end, when I saw you and Susannah walking down the road toward me, at first I thought that there were seven of you. It seemed that you were accompanied by the guides that I traveled with on my other journeys—Casey, Reverend Tyndall, Gabriella, Jake, and Janey—but they were ghostlike people floating above you, then they disappeared. Does that jibe with your memory?"

Leah looked softly at Mitch through her brown eyes. "Casey, Jake, and Janey were there with us. From the beginning, they were our time-travel guides—or should I say 'our travel guides to those other dimensions'? Yet they were only semi-real to us, guiding us along on our quest to reunite with you. It is as if they were in a bubble above us. Then, Susannah and I felt another presence looking over us, the same way that you felt my presence on the second night's journeys. The man was strangely dressed with a string tie, a priest's collar, and cowboy boots. At the end, a fifth presence hovered over us. The woman was striking, as she had gold irises and long blond hair."

Mitch recalled hearing this before at the end of their journey in Door # 7. They sat in silence together.

"Mitch, could all of this been more than a dream? It all felt so real. I can still feel the sting of the tears in my eyes when we worked together with Leah-Ann, our great granddaughter times-six. Could this all be real—were we in another dimension of the universe? Are we still back there in some way?"

A knock on their bedroom door got their attention.

"Come in, sweetie," said Leah.

Susannah entered and jumped into their bed.

"Mom, Dad, you won't believe this dream that I just had. You were both in it and we were riding through the future and—"

"Slow down, Susannah," said Mitch. "We all had the same dream."

She looked frightened. "How could that be? You can't look inside my head? You must be mistaken, Dad."

Leah smiled at her daughter. "It's okay, sweetie. You were with me on a night long journey trying to follow Dad. Try not to wrap your head around it too much. Just Let it Be."

Susannah quipped, "Let it Be! Let it Be! It seems like that is all I heard during that dream last night. I am not going to let you hide behind that. You both seem to know what is going on and I only have a snapshot glimpse of these other 'dimensions,' as you called them. I deserve to know what I was involved in, at the very least so that I can be prepared if it happens again!"

Mitch stroked Susannah's long blond hair. "Well, honey, first of all you will not experience this shared dream again . . . it was our final journey. On the first journey, I was on my own. Do you remember a few months ago when I was acting a bit weird?"

Susannah nodded her head. "A little weird? My God, Dad, you had us scared to death."

Mitch smiled at his daughter. "Yeah, well that was after my first all-night journey." He proceeded to share the full details of his trips through Doors 1, 2, and 3.

He concluded, "I believe I was supposed to learn about the choices that we have in changing society, specifically toward climate change. Every increase in our effort to deal with climate change gave us a better picture of what kind of world we will be giving to your generation and your children's generation."

Leah took the baton from Mitch. "I can tell you about the second night's journey, since I was along for the ride. I followed Dad as he entered Door #4, Door #5 and Door #6. I was more or less suspended above him, just as you and I were last night, though he could not see me. I was an observer of his travels."

Mitch interjected, "Well, I was aware of something watching me, but yes I did not know that it was Mom."

Leah continued, "The lesson that Dad learned from the second night of his time/place travels through the next three dimensions was to be open minded. We all have our beliefs about different things and we believe that we are absolutely correct. Do you understand that?"

Susannah nodded. "Kind of . . . Like when I had the argument with Janice at school about the best plays that we could do on our basketball team. I thought that she was so wrong!"

Leah responded, "But if you discussed this calmly with Janice, do you think that you would 'give in' and lose what you believed in?"

Susannah thought about this. "Well, maybe not. If I learned why she thought that a zone defense would be better for the team, I could re-evaluate why I thought that our man-to-man defense was better. And if I still felt strongly about that, I could calmly argue why I felt that I was right. But what if I was persuaded that she was right? Wouldn't I be giving in and losing the argument?"

Mitch answered this. "No. You would be growing in your perspective. You would gain from the knowledge that Janice gave to you."

Leah then turned to the third night. "So, Susannah, last night we were following Dad's trail and we picked up information from him as we moved from where he initially landed in 2034 and then tracked his journeys in 2152 to learn what he was doing in Buffalo, Nebraska, and Nigeria. What did we learn from following him through Door 7?"

Susannah thought for a while. "Well, the world could have a very different way of dealing with the climate problem. Instead of relying on government to lead the way, people took the problem into their own hands. The good things that the grassroots people started in 2034 spread like wildfire and every community offered some part of the solution. There was resistance throughout, causing the deaths of Reverend Tyndall and Gabriella. But the work of each of these groups was in synch like the members of a crew team. Those who fought the flow couldn't target whole governments, as the power to make these changes was diffused through different communities. The more success the local people made, the more attractive it was to bring others into the fold."

Mitch and Leah looked at each other in surprise. He spoke for both of them. "Well, Susannah, you hit the nail on the head and I am proud of how you grasped the message."

She looked at her father. "Dad, I want to say that I was very proud of you for the impact that you made on your various stops, but that is where I get confused. Did you make an impact on the futures that you visited or was the dream just a way of teaching you lessons from your observations on the different ways that we can deal with climate change here and now?"

Leah responded, "I suppose that we will never know the answer to that question."

Mitch countered, "Here is a thought that will get to the fundamental question we are all asking ourselves: Are the three of us still part of the farm in Nebraska and the Albany community of 2152? Let's all close our eyes and take a few minutes to think about the 2152 world we left."

After a five-minute silence, the three of them came out of their trance.

Susannah broke the silence. "Wow! I could feel and almost see the three of us playing our separate roles as if a copy of us may be still back there." Mitch and Leah looked at each other, silently agreeing.

Mitch spoke, "That is what scares me. If I think about that for long enough, I might be drawn into the same funk that I was in after my first night's journey."

Susannah gave Mitch and Leah a profound answer. "That is easy, Dad. It's what you were constantly telling people and being told by them throughout your journey . . . Let it Be."

Final Endnote of Book:

Climate change is a serious challenge to our society. As we process the journeys of this book, we can continue to ask ourselves questions:

- ***Is climate change real?*** Behind Door #4, we dealt with the possibility that there may not be enough scientific evidence and that the climate may not behave as predicted. However, using the scientific method, we weighed evidence and at this time the evidence is overwhelming that our earth is warming, that we are the cause, and that the consequences will be dire if nothing is done. Yet, we can still consider the very small possibility that science's current thinking may not be precisely correct.

- **If we believe the current evidence, are we doomed?** The journeys through Doors 2, 3, 5, 6 and 7 show us that there is still time to reverse our current trajectory toward an overly rapid change of climate – a "climate tipping point" that will wreck havoc on society. Yet, the time is running out.

- **Are there other solutions beyond conventional mitigation to solve the climate crisis?** Of course there are. Doors 5 and 7 present two scenarios. We also observed the AARTs—the Ambient Air Removal Technology, which is itself geo-engineering. Geo-engineering remains the "unspoken wrong path" among most of us who know about it. But who is to say that there may not be a circumstance where we use geoengineering to cool the Earth's temperature artificially reducing the temperatures to reverse the society-wide disaster if the "too late to mitigate" moment comes upon us? Yes, a *temporary* use of geo-engineering may be one of the tools to have in our pockets in case our backs are against the wall of devastation. Yet, the key word is "temporary." Just as Dr. Mitch of Door #5 doubted that we could ever wean ourselves away from the temporary solution, we would need to be just as skeptical. We would need to use this as a last resort to give us a second chance at real mitigation and stop the tinkering with our heat source as soon as possible. We would need to set a short window of time—say, ten years—and then dismantle the geo-engineering. During this short period of time, we would need to phase in emission controls to reduce greenhouse gas emissions by 80 percent. Would we really be able to do this?

- **Why is there a door (#6) with a volcano that cools the earth? What does this have to do with global warming?** First of all, cooling *is* climate change. Many skeptics point to the fact that with all of the worrying that people do about global warming, one eruption from a supervolcano would make it all irrelevant. Secondly, if society does survive and industrialization re-emerges, we will need to adapt to a world that will return to pre-eruption conditions. Can we adapt with wisdom? That would be our challenge.

- **Is the utopia of Door #7 just a fantasy?** We'll never know unless we try. As indicated in the endnotes to the previous chapter, there are already sustainable grassroots organizations emerging in local communities to create solutions of their own. If it grew into an international movement, the momentum might be the path that gets us to the solution in a manner that cuts through the current political divisions that we have today on government-mandated mitigation. Using the power of the people in the local communities with coalitions of industries, utilities, new technologies, religious institutions, philanthropists, and concerned citizens, we may be able to work to the solution from the bottom up.

- **So, which door should we enter?** Though the authors have their own personal preferences, we do not endorse them here. It is up to the reader to decide. What we do endorse is that every citizen become informed and involved in an open-minded national and international debate about whether we have a climate emergency problem and, if so, what to do about it., Then, we as citizens should take the bull by the horns and take action to do what we need to do. That way, we can dispose of our worries and take the "Let it Be" attitude, as we are taking action and involved in the battle against climate change.

What worries the authors the most is the attitude that "we have made our bed, so we must lie in it. We screwed up, so we are doomed." Despite the fact that most people believe that climate change is real, we think that many people have given up. In this case, the theme of "Let it Be" from our journeys of this novel does not mean surrendering . . . giving up means that we are writing off the world of our children and grandchildren. Even if we fail to ward off climate change, at least we will have given it our best try. In December, 2015, the world came together in Paris and agreed to a global accord to reduce greenhouse-gas emissions to levels that would stabilize the world's temperature to 2 degrees Celsius increase from pre-industrial times (the accord was signed in April 2016). It is a weak treaty, because it is non-binding. Yet, it was the world's first baby step in the right direction. At the very least, however, all nations on Earth now recognize the need to combat climate change with specific actions. We often hear the saying that "a journey of a thousand miles begins with a single step." The *second* step looms ahead: to achieve serious emission reduction targets. So, let's roll up our sleeves—with open minds—and get to work.

ACKNOWLEDGMENTS OF THE PRINCIPAL AUTHOR

This book has been a labor of love that transcended my contributions to its production. It has been a team project. I gratefully acknowledge each member of the Team:

- Dr. Harry Rosvalgi, the STEM Director of the Danbury, Connecticut School System, who gave me early feedback and encouragement to move forward with the writing of the book.
- All of the editors, co-editors, and proofreaders who worked tirelessly to shape the writing from rough into fine.
- My co-author, Brian, who integrated color into the characters and the plot and contributed his vision of future settings of the seven doors.
- Paul Fargis, a retired publishing professional, who gave me key advice every step of the way on the process for molding the rough story into a publishable book.
- To my friends John Fransence and Mark Ligorski, and the late Dennis Elpern, who reviewed an early version of the manuscript and provided valuable feedback.
- To my dear Dad, who has been a role model to me throughout my life. He represents to me the quintessential personification of the "Let it Be" quote throughout the novel.
- Vice President Al Gore, who I look up to as an enlightened visionary who raised the issue of climate change in the 1980s into public consciousness and who has tirelessly worked to educate the public on the action needed to take on the challenge of protecting the Earth for future generations.
- And there is Dr. Mitch. Does he exist? Perhaps he is a climate ecologist who inspired me to start this work.
- Finally, to my wife Nancy who has stood by my side from the get go. She has been my most constructive critic and advised me on every stage of the development of the book with substantive recommendations. She also proofread the entire work several times, catching errors that others missed.

Thank you all.

www.ingramcontent.com/pod-product-compliance
Lightning Source LLC
Chambersburg PA
CBHW080652190526
45169CB00006B/2087